S0-ADV-167

The various steps in the manufacture of a high-pressure sodium lamp as seen through the highly imaginative eyes of J. Eijsermans, a development engineer.

THE HIGH-PRESSURE SODIUM LAMP

THE HIGH-PRESSURE SODIUM LAMP

dr ir J.J. de Groot
and
ir J.A.J.M. van Vliet

MACMILLAN EDUCATION

PHILIPS TECHNICAL LIBRARY
KLUWER TECHNISCHE BOEKEN B.V. -
DEVENTER - ANTWERPEN

First published in the United Kingdom 1986 by
MACMILLAN EDUCATION LTD
Houndmills, Basingstoke, Hampshire RG21 2XS
and London

ISBN 0 333 43245 2 (Macmillan Edition)
ISBN 90 201 1902 8 (Kluwer Edition)

© 1986 Kluwer Technische Boeken B.V. - Deventer

1e druk 1986

Niets uit deze uitgave mag worden verveelvoudigd en/of openbaar gemaakt door middel van druk, fotokopie, microfilm of op welke andere wijze ook, zonder voorafgaande schriftelijke toestemming van de uitgever.

No part of this book may be reproduced in any form, by print, photoprint, microfilm or any other means without written permission from the publisher.

Preface

The purpose of this book is to outline the principles on which the design of high-pressure sodium discharge lamps is based and to leave the reader with a better understanding of both the background to and the application of these principles.

The book is addressed to research workers, students and development engineers engaged in the study of gas discharges in general and high-pressure sodium discharges in particular.

The emphasis throughout is on the physical aspects of the discharge, concentrating in the main on work carried out in the seventies and the early eighties. This work followed on from the early development work on lamp materials that took place in the beginning of the sixties. The renewed interest in the chemical and technological aspects of the high-pressure sodium discharge lamp coincides with the current development of low-wattage and improved-colour lamps. Because these aspects are still in the development stage, they have been covered somewhat less extensively. An exhaustive treatment of the applications open to the high-pressure sodium lamp is outside the scope of this book.

Chapter 1 provides an introduction to the high-pressure sodium lamp and examines its most important basic features in comparison with other discharge lamps.

Chapter 2 deals in general terms with the power balances and spectra of typical high-pressure sodium lamps and so provides an insight into the energy conversion and the luminous efficacy aspects of these lamps.

High-pressure sodium lamps are generally based on a discharge in a mixture of sodium, mercury and xenon, and this makes a theoretical description rather difficult. To simplify matters, therefore, a high-pressure sodium discharge without these buffer gases is first considered. The spectrum and the power balance of such a so-called pure sodium discharge are discussed in more detail in Chapter 3 and 4 respectively.

In Chapter 5 attention is centred on the influence that the buffer gases mercury and xenon have on the field strength, plasma temperature, radiant power and spectrum of a high-pressure sodium discharge.

Chapter 6 deals with the ignition of the HPS lamp and its stabilisation on the usual 50 Hz mains-frequency supply. The special cases for high-frequency and pulse current operation are treated in Chapter 7.

Chapters 8 and 9 examine the general properties of the discharge-tube material, the ceramic-to-metal seal and the electrodes.

Finally, Chapter 10 provides guidance on lamp design, the material presented in the earlier chapters being used to shape a design approach.

The authors are indebted to Dr M. Koedam, director of the Lighting Division within the Nederlandse Philips Bedrijven B.V. in Eindhoven, for his support in making the publication of this book possible. We also wish to thank Prof. Dr A.A. Kruithof for his careful reading of the manuscript, and many of our colleagues in the Philips Lighting Division and the Company's Research Laboratories for their invaluable comments. Finally, our thanks are also due to Mr A. Kortman, who produced the many drawings, and to Mr D.L. Parker for his invaluable assistance in preparing the manuscript.

J.J. de Groot   Eindhoven
J.A.J.M. van Vliet   May 1985

Contents

List of colour plates

Chapter 1

Introduction

1.1 Historical Review

1.1.1 History of Discharge Lamps

Mankind has been making use of artificial light for thousands of years. But it was only towards the end of the last century that the role played by sodium in both daylight and artificial light was discovered. In 1860, Kirchhoff showed that the reversed D-lines in the spectrum of the sun observed by Wollaston and Fraunhofer are identical to the yellow sodium D-lines in the light emitted by flames.

Artificially contrived gas discharges date back some three centuries, to the time that methods for removing the air from a vessel were invented. The term discharge originated in the time of the first experiments with current conduction through a gas due to the discharge of a capacitor. One of the first, accidentally-caused gas discharges was that observed by Pacard one evening in Paris in 1676 when carrying a mercury barometer. The movement of the mercury in the Torricelli 'vacuum' caused a light phenomenon to occur. In 1742 Christian August Hansen experimented with a vacuum tube containing a small amount of mercury and found that on applying a high d.c. voltage the tube emitted light. Better known are the Geissler low-pressure discharges (from around 1856) in glass tubes 'evacuated' with a mercury pump and operated on a high-voltage a.c. current supply. The period 1890–1910 witnessed the invention of a whole family of low-pressure as well as high-pressure mercury discharges as possible light sources. But is was not until around 1920 that a discharge in sodium vapour at low pressure ($\approx 0.5\,\text{Pa}$) was obtained, since this had to await the development of a sodium-resistant glass (Compton and Van Voorhis, 1923; Compton, 1926, 1931).

It was anticipated that the colour appearance of the monochromatic yellow light emitted by the low-pressure sodium (LPS) lamp (colour plate 1 spectrum a) would be improved with little loss in luminous efficacy if the sodium vapour pressure could be increased by a factor of between 10 000 and 100 000. A theoretical assessment in 1958 supposed that the improvement in colour of the resultant high-pressure sodium (HPS) lamp would come from the enhancement of the non-resonant sodium lines relative to the resonance lines, just as for mercury (Cayless, 1982). What was not foreseen was the paramount importance of the strong broadening and self-reversal of the sodium D-lines occurring at high sodium vapour pressures (colour plate 1 spectra b and c).

The main problem that had to be solved to obtain an HPS lamp was to find a light-transmitting discharge tube material resistant to the highly reactive sodium at the high temperatures involved. This problem was studied almost simultaneously in Great Britain and in the United States. One of the early studies of HPS discharges was made in 1959 by Clarke and Moore (Cayless, 1982). Their HPS 'lamp' is shown in figure 1.1 (see also ref.: Light & Lighting, 1966). The discharge vessel was made of 'Nilo K' pieces welded together.

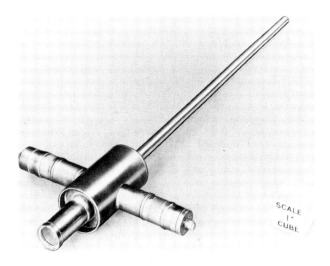

Figure 1.1 High-pressure sodium discharge vessel made in 1959 by Clarke and Moore.
The arc ran between tungsten electrodes mounted on spark plug insulators in the side arms, the light being emitted through the circular sapphire window. The nickel tube opposite this window controlled the sodium vapour pressure.

The arc discharge took place between thoriated tungsten electrodes mounted on 'Regalox' (86% alumina) spark-plug insulators in the side arms. The discharge was visible through the circular sapphire window sealed in the metal body. The long tube opposite the window was for exhausting the vessel and dosing with sodium. It was made of nickel and also functioned as a cool tip. The vessel contained sodium and about 650 Pa of argon. It was operated in an oven at about 900 K with a current of 1 to 2 A.

1.1.2 First High-Pressure Sodium Lamp

The first practical HPS lamp was constructed in the United States in the beginning of the sixties. An important preliminary step in the development of this lamp took place between 1955 and 1957 when Cahoon and Christensen (1955), followed by Coble (1962), developed a method for obtaining translucent, gas-tight polycrystalline alumina (see also Matheson, 1963). This involved sintering 99.9 per cent pure Al_2O_3 with the addition of about 0.1 per cent of MgO, at a temperature of about 2100 K. This amount of magnesia was added to get optimum light transmission (figure 1.2).

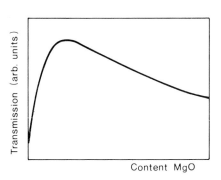

Figure 1.2 Light transmission of sintered alumina as a function of the magnesia content.
The transmission is optimum for a magnesia content of 0.05-0.1 per cent by weight, depending on the wavelength considered. (Coble, 1962)

The next important step was to make a vacuum-tight, sodium-resistant connection between the electrical feedthrough and the discharge tube. Stainless-steel tubes were used for the electrical feedthroughs connecting the tungsten electrodes and the current supply (Schmidt, 1961; Louden and Schmidt, 1965; Schmidt, 1966). These tubes were brazed to metal caps consisting of a nickel-chromium-iron alloy, having a high melting point and a coefficient of expansion close to that of alumina. These metal caps were brazed to the ends of the alumina discharge tube using thin titanium washers to metallise the ends of the ceramic tube. One of the stainless-steel tubes was used to evacuate the discharge tube prior to introducing the sodium and the starting gas.

The first laboratory lamps were made in about 1958 (figures 1.3 and 1.4). For reasons of simplicity, these first lamps employed xenon and caesium discharges (General Electric, 1961; Schmidt, 1961; Louden and Schmidt, 1962; McGowan, 1976; McVey, 1980). As expected, of the various alkaline metals, sodium proved to be the most promising for making an efficient light source. With it, luminous efficacies up to 145 1m W^{-1} were obtained if xenon at high pressure was used as the buffer gas (Schmidt, 1963b and 1966). However, the high starting voltage needed to ignite a lamp having such a high xenon pressure made this approach impracticable at that time.

15

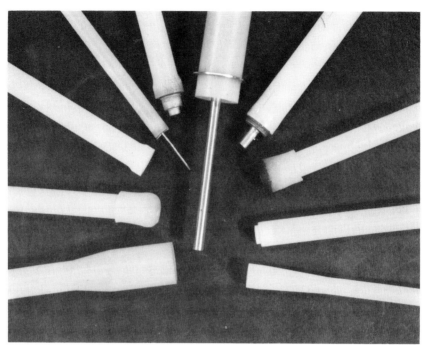

Figure 1.3 The first polycrystalline alumina discharge tube (from around 1958) used for an xenon lamp, bracketed by a selection of later experimental arc tubes. (McVey, 1980)

Figure 1.4 Historical end-structures for high-pressure sodium (HPS) lamps. (McVey, 1980)

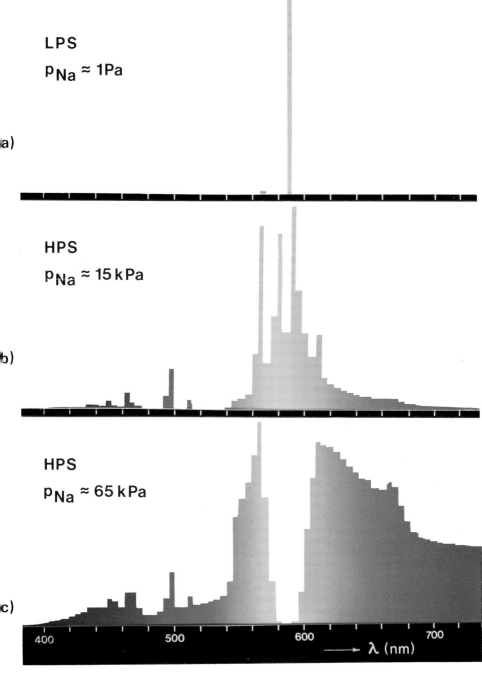

Plate 1 Visible spectra of (a) low-pressure sodium (LPS) lamp (b) standard high-pressure sodium (HPS) lamp and (c) 'white' HPS lamp with good colour rendering properties, showing the influence of the sodium vapour pressure (p_{Na}) on the spectrum. The spectra have been measured with integration intervals of 5 nm. For the HPS lamp the sodium D-lines at 589.0 and 589.6 nm are strongly self-reversed.

a)

b)

c)

Plate 2 Photographs of the same road under identical conditions and lighted to approximately the same level with:
(a)   phosphor-coated   high-pressure mercury lamps (HPL-N 400W)
(b) low-pressure sodium lamps (SOX 180W) and
(c) high-pressure sodium lamps (SON 250W).
These photographs give an impression of the differences in colour appearance of the various lamps employed. They were taken at the Philips open-air laboratory in Acht near Eindhoven, The Netherlands.

The first commercially available HPS lamps appeared in around 1965, almost concurrently in the United States, Great Britain and The Netherlands (Louden and Schmidt, 1965; Ruff, 1963; Nelson, 1966; de Vrijer, 1965). The first type was a 400 W version with a luminous efficacy of about 100 lm W^{-1}.

1.1.3 Progress in Recent Years

Continuous research and development since the introduction of the first HPS lamp have resulted in lamps with improved luminous efficacies (figure 1.5), a wide range of lamp powers and diversification in lamp design (Sec. 1.2). The HPS lamp is relatively new compared with most other lamp types (figure 1.5). However, its relatively high luminous efficacy and pleasant colour appearance have already led to its widespread application in rapidly-increasing numbers.

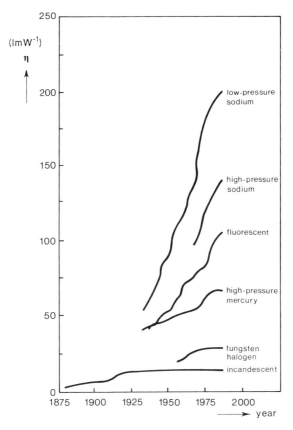

Figure 1.5 The luminous efficacies η of various lamp types over the last century.
The lamp power is not the same for the various curves. For the discharge lamps the ballast losses are not included.

17

Apart from the technological development of the HPS lamp, there has at the same time been considerable progress in understanding the physical and chemical processes occurring in this lamp. Improved or new measuring techniques and various computer models have helped to make the HPS lamp one of the best understood discharge lamps. Various aspects of the HPS lamp have been reviewed in recent years by de Groot *et al.* (1975a), McVey (1980), Wharmby (1980), van Vliet and de Groot (1981), Zollweg (1982) and Akutsu (1984).

1.2 General Aspects

The sodium lamp is based on the principle of an electric discharge in sodium vapour. The electrons, which gain their energy from the electric field existing between the electrodes, excite the sodium atoms, which then emit the yellow D-lines and other characteristic sodium lines. For the LPS lamp the sodium vapour pressure is about 0.5 Pa (Denneman, 1981), while for the HPS lamp a sodium vapour pressure of 10 kPa or higher is needed.

Sodium, the most important physical properties of which are given in table 1.1, is a soft, silver-white, highly-active, low-melting-point metal. The discharge takes place in a light-transmitting discharge tube made from a material that is able to withstand the reactive sodium at temperatures above 1000 K.

Table 1.1 Some important physical properties of sodium (Na) and mercury (Hg)

	Na	Hg
Atomic number	11	80
Atomic weight	23.0	200.6
Melting point	370 K (97 °C)	234 K (−39 °C)
Boiling point	1163 K (890 °C)	630 K (357 °C)
Energy of lowest excited level	2.10 eV	4.89 eV*
Important resonance line(s)	589.0/589.6 nm	253.7 nm
Oscillator strengths of these lines	0.66/0.33	0.02
Ionisation potential	5.14 eV	10.43 eV

* This value relates to the $3P_1$ level, which is the lowest non-metastable level of the Hg atom.

Polycrystalline alumina is the material normally used. A typical HPS discharge tube is shown in figure 1.6. In this case the electrical feedthrough, in the form of a niobium tube, is sealed into the alumina tube using a sealing ceramic. Typical complete HPS lamps are shown in figure 1.7.

The HPS lamp is similar in several respects to the well-known high-pressure mercury vapour (HPMV) discharge lamp (see e.g. Elenbaas, 1951, 1965;

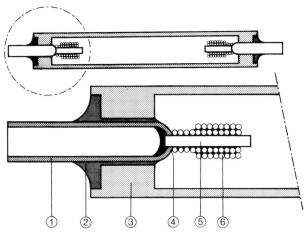

Figure 1.6 Typical discharge tube in an HPS lamp. The niobium tube, carrying the tungsten electrode, is sealed into the alumina tube using a sealing ceramic.
1 niobium tube current feedthrough
2 sealing ceramic
3 alumina
4 brazing metal (Ti)
5 tungsten rod
6 tungsten windings with emitter

Figure 1.7 Photograph of typical high-pressure sodium lamps (SON 400W and SON/T 400W).
The discharge tube is mounted in an evacuated outer bulb, which may be coated ovoid (left) or
clear tubular (T) (right).

19

Waymouth, 1971). At the early development stage it was hoped that the HPS lamp would combine the high luminous efficacy of the LPS lamp and the acceptable colour rendering properties and compactness of the HPMV lamp. These ary very important features* for a lamp. Unfortunately, the goals of high luminous efficacy and good colour rendering are often mutually conflicting, and may together make it difficult to satisfy a third design goal, important for any lamp, namely a long lifetime. Thus, as with most practical lamps, the present HPS lamp is a design compromise between the various ideals mentioned. Still, it was just what was wanted to fill the 'gap' in the existing lamp range between those lamps with a high luminous efficacy but poor colour rendering properties and large volume, such as the LPS lamp, and those lamps with reasonable colour rendering and small volume but with a relatively low luminous efficacy, such as the HPMV lamp (Sec. 1.2.3).

1.2.1 Comparison Between Sodium and Mercury Discharges

Sodium and mercury discharges are in many aspects similar. Both discharges reveal two peaks in the curve representing the luminous efficacy as a function of the vapour pressure (figure 1.8a). One peak occurs at a pressure of 0.5–1 Pa for the LPS and low-pressure mercury (fluorescent) lamps and one at a pressure of 10^4–10^5 Pa for the HPS and HPMV lamps. Both high-pressure lamps, with their high current density and high input power per unit discharge

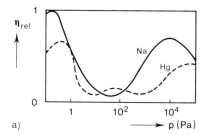
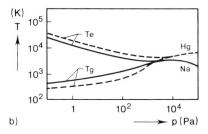

Figure 1.8 Comparison between sodium and mercury discharges showing (a) the relative luminous efficacies η_{rel} of sodium and mercury discharge lamps and (b) the electron and gas temperatures (T_e and T_g respectively) at the axis of the discharge as functions of the sodium or mercury vapour pressure (p).
A fluorescent powder is used for the mercury discharge lamps to convert the ultraviolet radiation into light. Data in (a) from British Lighting Industries (ref: Light & Lighting, 1966), data for mercury in (b) from Elenbaas (1951).

---

* The luminous efficacy η gives the luminous flux per unit electric input power, while the general colour rendering index R_a is a number (maximum value is 100) denoting the relative ability of the lamp to render objects illuminated by it in their natural colours (CIE, 1974). And, of course, the more compact a lamp is, the easier it is to design efficient luminaires having the light distribution needed for a specific application.

length (table 1.2), have a high gas temperature at the axis of the discharge, the gas temperature being approximately equal to the electron temperature (figure 1.8b). These high-pressure discharges are so-called thermal arcs. They show a steep temperature gradient between the hot core at the axis and the wall. In contrast with the high-pressure discharges, both low-pressure discharges, with their low current density and low input power per unit discharge length (table 1.2) are non-thermal arcs; they have a low gas temperature and a high electron temperature (figure 1.8b). Both these temperatures are nearly homogeneously distributed over the whole discharge volume.

Table 1.2 Some characteristics of typical high-pressure and low-pressure sodium and mercury discharge lamps

Type		high-pressure sodium SON/T	high-pressure mercury* HPL de Luxe	low-pressure sodium SOX-E	low-pressure mercury* TLD/84
Lamp power (P_{la})	(W)	400	400	130	36
Discharge tube material		alumina	fused silica	soda-lime glass + borate layer	soda-lime glass
Discharge tube inner diameter	(mm)	7.5	19	19	24
Electrode spacing	(mm)	82	72	1890	1120
Gas/vapour pressures (in operating lamp):					
Na	(Pa)	$1\ 10^4$	–	0.5	–
Hg	(Pa)	$8\ 10^4$	$4\ 10^5$	–	1
Noble gas pressure (Ar, Kr, Ne, Xe)	(Pa)	$2\ 10^4$	$2\ 10^4$	730	200
Lamp current	(A)	4.4	3.2	0.62	0.44
Lamp voltage	(V)	105	140	245	102
Discharge input power per unit of length	(W m^{-1})	4500	5500	60	30
Electron density at axis	(m^{-3})	10^{21}–10^{22}	10^{21}–10^{22}	10^{18}–10^{19}	10^{18}–10^{19}
Gas temperature at axis	(K)	4000	6000	530	320
Electron temperature at axis	(K)	4000	6000	10000	13000

Table 1.2 (Continued)

Type		high-pressure sodium SON/T	high-pressure mercury* HPL de Luxe	low-pressure sodium SOX-E	low-pressure mercury* TLD/84
Degree of ionisation of the metal at axis	(%)	4	0.02	50	1.2
Discharge tube wall temperature	(K)	1500	1000	530	320
Efficiency for radiant power (P_{rad}/P_{la})	(%)	56	48	44	66
Efficiency for visible radiation (P_{vis}/P_{la})	(%)	31	17	39	28
Luminous efficiency of visible radiation		0.57	0.51	0.76	0.48
Luminous efficacy	(lm W^{-1})	120	60*	200	92
Colour appearance		golden-white	white	yellow	white
Colour rendering index		23	47	–	86

* For low-pressure mercury (fluorescent) lamps and for most high-pressure mercury lamps a fluorescent powder is used to convert the ultraviolet radiation into light.

A more detailed comparison also shows characteristic differences between sodium and mercury discharges (see also tables 1.1 and 1.2)

a) Sodium is more reactive than mercury. Furthermore, the temperature needed to obtain the same vapour pressure is higher for sodium than for mercury. So the requirements placed on the discharge tube material with respect to thermal and chemical stability are much more severe for sodium than for mercury.

b) The strongest sodium resonance lines lie in the visible part of the spectrum, whereas the mercury resonance lines lie in the ultraviolet region. In most mercury discharge lamps therefore a fluorescent powder is used to convert ultraviolet radiation into light in order to obtain the maximum luminous flux.

c) The oscillator strengths of the most important sodium resonance lines (589.0/589.6 nm) are much larger than that of the most important mercury resonance line (253.7 nm). This explains the stronger (resonance) broadening and the increased self-absorption of the former lines as compared to

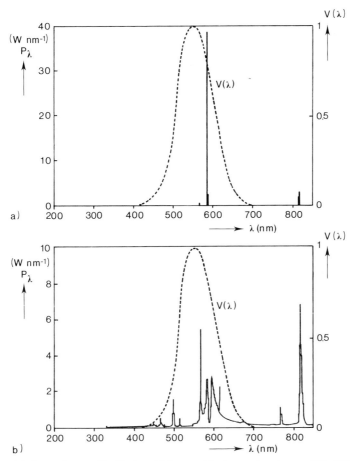

Figure 1.9 Spectral power distributions (full curves) in the wavelength range 200–850 nm of (a) a low-pressure sodium lamp (SOX-E 131) and (b) a high-pressure sodium lamp (SON 400 W). V(λ) (dashed curve) indicates the spectral luminous efficiency for the human eye as a function of the wavelength.

the latter (see figures 1.9 and 1.10), as the resonance broadening and self-absorption are both proportional to the oscillator strength of the line considered. Because of the strong self-absorption of the sodium D-lines, this phenomenon strongly influences the visible spectrum of the HPS lamp, especially at increasing sodium vapour pressure (colour plate 1). The characteristic differences in the spectra of sodium and mercury discharges at increasing vapour pressure may also explain why the peaks in the luminous efficacy appear at a lower vapour pressure for the sodium discharge lamp than for the mercury discharge lamp.

d) The ionisation energy of sodium is much lower than that of mercury, so the (electron) temperature will be lower in a sodium discharge than in

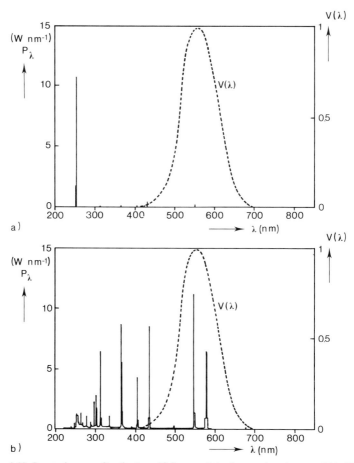

Figure 1.10 Spectral power distributions (full curves) in the wavelength range 200–850 nm of (a) a low-pressure mercury lamp with germicidal glass (TUV 40W) and (b) a high-pressure mercury lamp (400 W) with quartz outer bulb and without fluorescent powder. V(λ) (dashed curve) indicates the spectral luminous efficiency for the human eye as a function of the wavelength.

a mercury discharge (despite the lower vapour pressure in the former), as the electron densities are about the same for both discharge types. A consequence of the lower vapour pressure in a sodium lamp is that the degree of ionisation at the axis of the sodium discharge is much higher than in the mercury discharge.

e) Another consequence of the relatively low sodium vapour pressure in a pure sodium high-pressure discharge (at maximum luminous efficacy) is that the electric field strength is lower than in an HPMV discharge lamp, both the electron densities in these discharges and the collision cross-sections for sodium and mercury being roughly the same. To increase the electric field strength in HPS lamps, narrow discharge tubes are used in combination with a buffer gas. Usually mercury is added to the discharge in the form of a sodium amalgam.

The efficiency with which the electric input power is converted into radiation (ultraviolet, visible and infrared) is roughly the same (about 50%) for HPS and HPMV lamps (table 1.2). Radiation in the visible wavelength region, however, is produced much more efficiently in the HPS lamp (31%) than in the HPMV lamp (17%) because of the characteristic differences between the spectra already mentioned. This, combined with the relatively high luminous efficiency of the visible radiation for the sodium spectrum (see also the luminous efficiency curve in figures 1.9 and 1.10), results in the HPS lamp having a luminous efficacy that is about twice that of the HPMV lamp.

Technologically the HPS lamp is quite different from the HPMV lamp. The well-known pinch seal as used in HPMV lamps cannot be used because the alumina does not soften when heated, as do glass and fused silica. Special sealing ceramics can be used for making the seal. The tungsten electrodes in both these high-pressure discharge lamps are in principle the same.

1.2.2 System Aspects

As with almost all discharge lamps, the HPS lamp must be operated in conjunction with a current-limiting ballast. With an a.c. supply the ballast is usually an inductance or choke placed in series with the lamp. The fact that the HPS lamp normally operates with saturated sodium and mercury vapour, makes the stabilisation of this lamp difficult. This is because the lamp voltage is sensitive to temperature variations at the coldest spot in the discharge tube, which result in variations in sodium and mercury vapour pressures. The lamp voltage is thus largely dependent on the lamp power and also on the thermal insulation (luminaire). For the HPMV lamp this is hardly the case, as all the mercury is vaporised.

Like all metal vapour lamps, the HPS lamp contains an ignition gas. This is because the sodium and mercury vapour pressures of the sodium amalgam at room temperature are too low ($< 10^{-5}$ Pa) to permit of ignition. In HPMV lamps the mercury vapour (pressure about 0.1 Pa at room temperature) in combination with the argon ignition gas, forms a Penning mixture, which results in a relatively low breakdown voltage. The use of xenon as the ignition gas in most HPS lamps, resulting in a relatively high breakdown voltage, necessitates the use of a starter. An example of starter (electronic ignitor) and ballast is shown in figure 1.11.

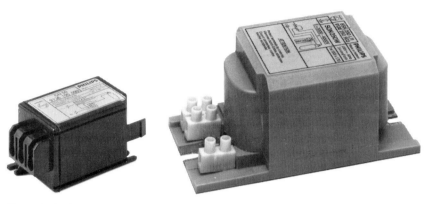

Figure 1.11 Photograph of electronic ignitor (left) and choke ballast (right) for a 250 W HPS lamp.

1.2.3 Light Quality and Luminous Efficacy

When the sodium vapour pressure in an HPS lamp is chosen for maximum luminous efficacy (standard HPS lamp) the colour rendering is moderate (general colour rendering index $R_a = 23$). The fact that the visible spectrum is greatly influenced by the sodium vapour pressure, means that it is possible to modify the colour rendering properties of the light emitted. As already mentioned, the visible part of the spectrum of the HPS lamp, like that of the LPS lamp, is dominated by the radiation of the sodium D-lines (figure 1.9). However, the D-line radiation of the HPS lamp is spread over a much wider wavelength range, mainly owing to the pressure broadening and the self-absorption of the resonance radiation, which together lead to the strong self-reversal observed. At high sodium densities the radiation near the centres of the D-lines, where the mean free path length of a photon is only about 0.1 μm, is very strongly absorbed. For wavelengths more than a few nano-metres distant from the line centre, the mean free path length of the photons is about equal to the discharge tube radius, so these photons have a high probability of escaping from the hot core without being absorbed.

The characteristic changes in the spectrum of an HPS lamp with increasing sodium vapour pressure can be followed by observing the spectrum after lamp ignition (figure 1.12). When started cold, the HPS lamp in fact first exhibits a noble gas discharge spectrum (usually of xenon, as this is used in most HPS lamps as ignition gas), which is followed after some ten seconds by the characteristic spectrum of a low-pressure sodium discharge. This even-tually develops into a high-pressure sodium discharge spectrum with increas-ing self-reversal of the sodium D-lines as the sodium vapour pressure in-creases, the wavelength difference $\Delta\lambda$ between the maxima of the self-reversed D-lines being a measure for the sodium vapour pressure.

26

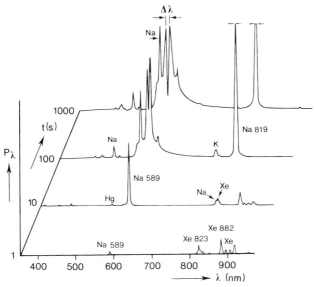

Figure 1.12 Development in time of the spectrum of a 150 W HPS lamp (with a high xenon pressure) after a cold start, as measured with an optical multichannel analyzer. A number of important lines are identified. t is the time, λ the wavelength and Δλ the wavelength separation between the maxima of the self-reversed sodium D-lines.

Because of the broadening of its spectrum, the standard HPS lamp has a 'golden-white' colour appearance and, in contrast to the monochromatic-yellow radiation of the LPS lamp, its light permits of some degree of discrimination between colours. Colour plate 2 gives an impression of the 'golden-white' appearance of the HPS lamp in comparison with the blueish-white colour of the HPMV lamp and the yellow colour of the LPS lamp.

The colour appearance of a lamp is objectively described by its chromaticity coordinates, which can be calculated from the spectral power distribution according to standard methods (CIE, 1974). In general it is desirable for the colour point of a light source to lie on or near the black body locus. The colour point of the standard HPS lamp lies very near the black body locus, as shown in the chromaticity diagram in colour plate 3a and figure 1.13. In addition to the sodium and mercury discharges, as discussed in the previous section, the colour points are given for the well-known incandescent and tungsten halogen lamps as well. The colour point given for a metal halide lamp refers to a lamp where, in addition to the iodide of sodium, thallium iodide and indium iodide are also added to the high-pressure mercury discharge. Also shown in colour plate 3a and figure 1.13 are the colour points of black body radiators at various temperatures. The (correlated) colour temperature T_c of a light source is defined as the temperature of the black body that emits radiation of the same chromaticity as the source considered. The

27

correlated colour temperature of a standard HPS lamp is about 2000 K. By increasing the sodium vapour pressure a 'white', 'super' or 'De Luxe' HPS lamp with higher correlated colour temperature can be obtained, approximating the colour point of an incandescent lamp, as is indicated by the arrows in colour plate 3a and figure 1.13.

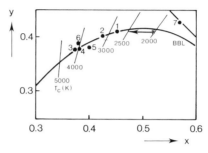

Figure 1.13 Enlarged part of the chromaticity diagram showing the black body locus (BBL) and lines of constant correlated colour temperature T_c. The colour point of a standard HPS lamp (SON 400 W) is indicated by an asterisk. The arrow indicates the shift in chromaticity coordinates (x,y) with increasing sodium vapour pressure. Further the colour points are given for various lamp types:
1. incandescent lamp (GLS 100W);
2. tungsten halogen lamp (1000W);
3. standard fluorescent lamp (TLD 36W/33);
4. three-band fluorescent lamp (TLD 36W/84);
5. phosphor-coated high-pressure mercury lamp (HPL de Luxe 400W);
6. metal halide lamp (HPI/T 400W);
7. low-pressure sodium lamp (SOX-E 131).

Improving the light quality of a lamp (as expressed in chromaticity coordinates, correlated colour temperature and colour rendering index) is mostly only possible at the expense of luminous efficacy. In general, for white lamps, the chromaticity coordinates of which lie near to or on the black body locus, a higher luminous efficacy can be obtained for lower values of the (correlated) colour temperature. This follows from the theoretical calculations of Mac-Adam (1950), presented in figure 1.14. For lamps with colour temperatures between 2000 K and 3000 K (as has the HPS lamp) the theoretical maximum luminous efficacy is 520 lm W^{-1}. In this case, all the input power has to be radiated in the visible part of the spectrum, with an optimum spectral power distribution. For practical discharge lamps the luminous efficacy is much lower than the theoretical maximum value because of various loss processes (electrode and conduction losses), the radiated power in the ultraviolet and infrared regions of the spectrum and the non-optimum spectral distribution of the visible radiation.

The wish for maximum luminous efficacy conflicts to a greater or lesser extent with the wish for good colour rendering. In general, for lamps with good

28

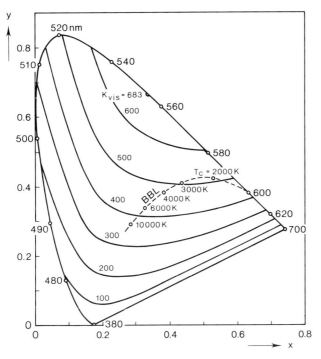

Figure 1.14 Chromaticity diagram with lines of constant maximum attainable luminous efficacy of visible radiation (K$_{vis}$). The black body locus (BBL) shows the colour points for black body radiators at various temperatures. On the spectrum locus (upper full curve), the colour points of various spectral colours are given. The spectral colours are characterised by their wavelengths given in nanometres. (MacAdam, 1950)

colour rendering properties, the visible radiant power has also to be spread over that part of the visible wavelength range where the luminous efficiency is relatively low. The relation between luminous efficacy and colour rendering index for the HPS lamp is given in figure 1.15 in comparison with that for various other lamps. It is obvious from this figure that the standard HPS lamp fills the 'gap' between the LPS lamp and the HPMV lamp: its luminous efficacy is much higher than that of the HPMV lamp, while its colour rendering capacity is better than that of the LPS lamp. The colour rendering index of the standard HPS lamp, which is less than that of the HPMV lamp, can be improved (up to a colour rendering index of 85) at the cost of luminous efficacy by raising the sodium vapour pressure (arrow A in figure 1.15).

The luminous efficacy of the standard HPS lamp can be improved by increasing the xenon pressure (arrow B in figure 1.15). This reduces the self-absorption of the sodium D-lines and the conduction losses, while the colour quality remains approximately the same. However, extra starting aids become necessary because of the higher breakdown voltage.

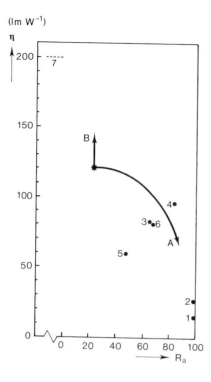

(lm W⁻¹) → η (lm W^{-1}) axis with values 200, 150, 100, 50, 0; R_a axis with values 0, 20, 40, 60, 80, 100.

Figure 1.15 Relation between luminous efficacy η and general colour rendering index R_a as found for various lamps. The standard HPS lamp (SON 400W) is indicated by an asterisk; the arrows A and B indicate the effects of raising the sodium vapour pressure or the xenon buffer gas pressure respectively. The numbers indicate the same lamps as given in figure 1.13. As the R_a-value has no meaning for the low-pressure sodium lamp (no 7), this lamp is indicated by a dashed line.

1.2.4 Applications

The standard HPS lamp is mostly employed for outdoor lighting. The lamp's high luminous efficacy and pleasant colour appearance, combined with its long (useful) life (between 10 000 and 20 000 hours) make it very suitable for all those applications where economy in operation is more important than colour rendering.

A survey of various outdoor applications with an indication of the lamp types commonly employed is given in figure 1.16. Good public lighting (viz. streets in residential areas, roads for motorised traffic and motorways) requires a range of luminous flux values between 2 and 70 klm, while for security lighting also lower and for industrial lighting and floodlighting also higher luminous fluxes are required. The present range of Philips HPS lamps, the main characteristics of which are given in table 1.3, covers the range of 3.5 to 125 klm and so, except in the low-lumen range, fulfils the requirements mentioned for public lighting. An example of road lighting with HPS lamps is shown in colour plate 3b (see also van Bommel and de Boer, 1981).

From an optical point of view the HPS lamp has much in common with the HPMV lamp. Both have a relatively small discharge tube in either a clear

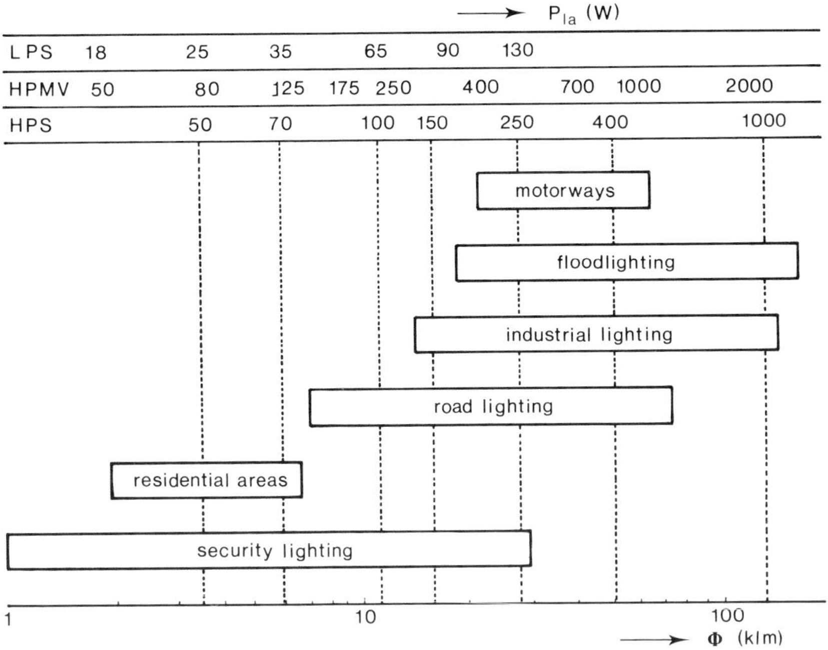

P_{la} (W)

L PS	18		25	35		65	90	130				
HPMV	50		80	125	175	250		400		700	1000	2000
HPS			50	70		100	150	250		400		1000

motorways

floodlighting

industrial lighting

road lighting

residential areas

security lighting

1   10   100

Φ (klm)

Figure 1.16 Survey of various lighting applications covered by various lamp types with their range of lamp powers. The bottom scale gives the required luminous flux.

or coated outer bulb (figure 1.7). The outer bulb may be coated on the inside with a layer of calcium pyrophosphate powder, which serves to enlarge the light-emitting area, so decreasing the lamp's luminance (table 1.3). The size and shape of the coated outer bulb are the same for both types of lamp, hence luminaires designed for the coated ovoid HPMV lamp can easily be equipped with the HPS lamp (see also figure 1.17). This results in a higher lighting level for the same installed power or, alternatively, a marked reduction in energy consumption – some 50 per cent – for the same lighting level. Special HPS lamps (SON-H) have been developed that can operate on ballasts designed for HPMV lamps; these lamps consume some 10 to 15 per cent less energy and provide 40 to 55 per cent more light than the HPMV lamps they replace.

HPS lamps with a clear (tubular) outer bulb can be used to advantage in those road lighting applications where the light has to be concentrated on the carriageway only. They are also extremely well suited for use in specially designed floodlight luminaires. The golden-white colour of the HPS lamp may be of great advantage for floodlighting tourist attractions (colour plate 4). The small diameter of the discharge tube permits of a high luminaire efficiency together with a sharp beam cut-off, the latter being important in connection with the avoidance of glare. As a result, an increasing number of

31

Table 1.3 Lamp current I_{la}, lamp voltage V_{la}, luminous flux Φ, average luminance L_v and luminous efficacies of lamp and system for present (1985) range of Philips high-pressure sodium lamps

Type	I_{la} (A)	V_{la} (V)	Φ (klm)	L_v (cd mm^{-2})	Luminous efficacy (lm W^{-1})	
					lamp	system
SON 50 W-E	0.76	85	3.3	0.04	70	60
SON 50W-I	0.76	85	3.3	0.04	70	60
SON-T 50W-E	0.75	86	4.0	3.0	80	69
SON 70 W-E	1.0	90	5.6	0.07	82	70
SON 70 W-I	1.0	90	5.6	0.07	82	70
SON-C 70 W-I	1.0	90	6	3.0	85	74
SON-T 70 W-E	1.0	86	6.5	3.0	90	78
SON 100 W	1.2	100	9.5	0.15	100	90
SON-T 100 W	1.2	100	10	3.0	105	94
SON 150 W	1.8	100	13.5	0.10	92	81
SON-S 150 W	1.8	100	15.5	0.12	102	91
SON-T 150 W	1.8	100	14	3.0	97	85
SON-ST 150 W	1.8	100	16	3.4	105	93
SON 250 W	3.0	100	25	0.19	100	89
SON-T 250 W	3.0	100	27	3.6	108	96
SON 400 W	4.4	105	47	0.24	118	108
SON-T 400 W	4.6	100	47	5.5	120	110
SON 1000 W	10.3	110	120	0.36	130	123
SON-T 1000 W	10.6	100	125	6.5	130	123
SON-H 210 W	2.5	104	18	0.14	86	78
SON-H 350 W	3.6	117	34.5	0.18	98	92

SON   indicates HPS lamp with coated ovoid outer bulb
SON-C indicates HPS lamp with clear ovoid outer bulb
SON-T indicates HPS lamp with clear tubular outer bulb
SON-H indicates HPS lamp to be used for direct replacement of the relevant high-pressure mercury lamp in existing installations.
The addition of the symbol S indicates a special version with a higher luminous efficacy (high xenon pressure), and the symbols E and I indicate lamps with external and internal ignitor respectively.

airport aprons, container terminals, sports training fields and the like are being lighted with HPS lamps housed in a suitable reflector-type luminaire (colour plate 5a). The standard HPS lamp may also be used in indoor lighting for those applications where colour rendering is not important, as may be the case for certain applications in the area of industrial lighting. Further the HPS lamp may be used for special applications, e.g. in lighting for floriculture (colour plate 5b) as the spectral power distribution of the HPS lamp coincides very well with the plant sensitivity curve for photosynthesis.

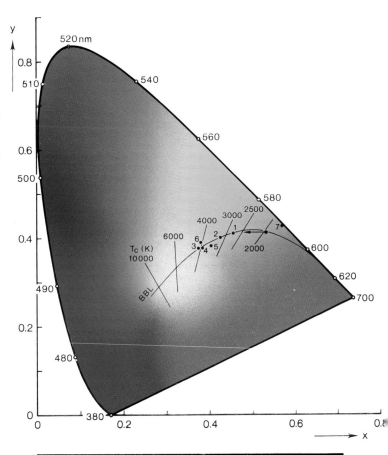

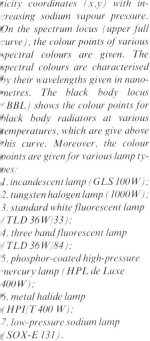

Plate 3 (a) The colour point of a high-pressure sodium lamp (SON 400W) is indicated by an asterisk in the chromaticity diagram. The arrow indicates the shift in chromaticity coordinates (x,y) with increasing sodium vapour pressure. On the spectrum locus (upper full curve), the colour points of various spectral colours are given. The spectral colours are characterised by their wavelengths given in nanometres. The black body locus (BBL) shows the colour points for black body radiators at various temperatures, which are give above this curve. Moreover, the colour points are given for various lamp types:

1. incandescent lamp (GLS 100W);
2. tungsten halogen lamp (1000W);
3. standard white fluorescent lamp (TLD 36W/33);
4. three band fluorescent lamp (TLD 36W/84);
5. phosphor-coated high-pressure mercury lamp (HPL de Luxe 400W);
6. metal halide lamp (HPI/T 400 W);
7. low-pressure sodium lamp (SOX-E 131).

Plate 3 (b) Example of road lighting with high-pressure sodium lamps (Barcelona, Spain).

Plate 4 Floodlighting of Brussels' city hall
with high-pressure sodium lamps. (Fischer, 1979)

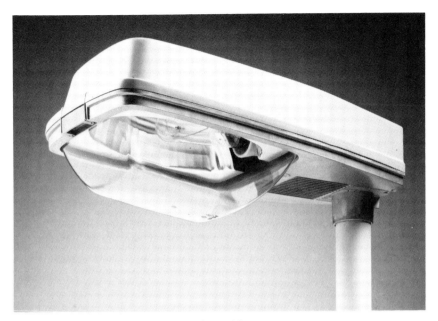

Figure 1.17 Typical road lighting luminaire for HPS lamps.

New types of HPS lamps with improved light quality as compared to that of the standard HPS lamp – white light and higher correlated colour temperature with good colour rendering properties – are being developed. At the relatively high sodium vapour pressures employed nearly all colours are sufficiently well represented in the spectrum (colour plate 1 spectrum c) to approximate the colour appearance and colour quality of an incandescent lamp (colour plate 7). This means that this lamp will become of importance for indoor lighting purposes, especially since the relatively small size of the discharge tube will permit of the construction of spotlights having even narrower beams than those possessed by the incandescent (bowl) reflector lamps, and with a significantly lower energy consumption.

Considering the above-mentioned features of the HPS lamp, it will be clear that it offers the possibility for attractive and economic solutions for many lighting applications.

Chapter 2

General Considerations Concerning Power Balance and Spectrum

Important lamp properties such as luminous flux and light quality are determined by the radiant power and its spectral distribution. To obtain an insight into the relevant processes that determine the amount of useful – in the case of HPS lamps this means visible – radiation, it is important to have an overview of the power balance in the lamp. The power balance describes the most important steps in the conversion of the electric input power into visible, infrared and ultraviolet radiation on the one hand, and into power dissipated by thermal and non-radiative loss processes on the other.

This chapter is devoted to some general considerations concerning the power balance and the spectrum, starting from measurements on practical HPS lamps. In Sec. 2.1 a qualitative description will be presented of the power flux in an HPS lamp. In Secs 2.2, 2.3 and 2.4 quantitative data will be given concerning the electrode power, the radiant power and the sum of conduction and absorption losses. In Sec. 2.5 the power balance will be considered for various lamp types. Sec. 2.6 will contain some final remarks on the subject of this chapter.

2.1 From Electric Power to Light

Of the total electric power fed to an HPS lamp, part is dissipated at the electrodes and part in the discharge (arc column)

$$P_{in} = P_{el} + P_d \tag{2.1}$$

where P_{in} = input power
P_{el} = electrode power
P_d = discharge (arc column) power

Of the discharge power, part leaves the discharge volume in the form of radiation and part is transferred by conduction to the discharge tube

$$P_d = P_{rad}^o + P_{con}^o \tag{2.2}$$

where P_{rad}^o is the radiant power from the discharge incident on the wall of the discharge tube and P_{con}^o is the conduction power at the inner surface of the discharge tube. The radiant power P_{rad}^o consists of ultraviolet (P_{uv}^o), visible (P_{vis}^o) and infrared (P_{ir}^o) radiation

$$P_{rad}^o = P_{uv}^o + P_{vis}^o + P_{ir}^o \tag{2.3}$$

Less than half of the electrode power is transferred to the discharge tube end construction, mainly by heat conduction along the electrode (some 40 per cent for a 400 W HPS lamp). The rest of the electrode power (about 60 per cent for a 400 W HPS lamp) is emitted as thermal (infrared) radiation.

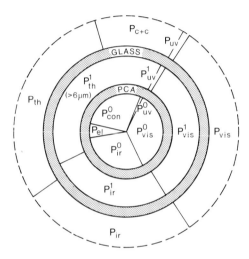

Figure 2.1 Schematic representation of the power flow in a 400 W HPS lamp.
Inside the PCA discharge tube the input power is converted into ultraviolet (P_{uv}^o), visible (P_{vis}^o) and infrared (P_{ir}^o) radiation, electrode power (P_{el}) and conduction power (P_{con}^o). The greater part of the discharge and electrode radiation passes through the alumina wall of the discharge tube (P_{uv}^1, P_{vis}^1, P_{ir}^1), the rest of the power leaving the discharge tube as thermal radiation (P_{th}^1). The greater part of the (near) ultraviolet, visible and infrared radiation passes through the glass outer bulb (P_{uv}, P_{vis}, P_{ir}), the rest of the power leaving the outer bulb as thermal radiation (P_{th}) and by way of conduction and convection (P_{c+c}).

At every physical boundary in the lamp (firstly the discharge tube, secondly the outer bulb) there is a transformation of energy. This is represented schematically in figure 2.1 for an HPS lamp, the discharge tube of which is placed in an evacuated outer bulb. As in Eqs (2.2) and (2.3), radiation and conduction in the discharge volume will be indicated (in the following) by a superscript 0. The greater part of the discharge radiation – viz. the radiation in the wavelength range 0.2 to about 6 μm where the alumina is translucent (see Sec. 8.1.3) – will pass through the discharge tube wall; the rest will be absorbed. Also the greater part of the thermal radiation of the tungsten elec-

35

trodes – with a maximum at about 1.5 μm for an electrode temperature of around 1700 K – will pass through the alumina wall of the discharge tube, the remainder being absorbed by this wall. The power absorbed by the tube wall, plus the power transferred to the wall by conduction – from the discharge or from the electrodes – can only leave the discharge tube in the form of thermal radiation in the case of an evacuated outer bulb, heat conduction through the support wires being negligible. The discharge and electrode radiation passing through the wall of the discharge tube will be indicated here by the superscript 1. At the outer bulb similar processes take place: the majority of the discharge and electrode radiation – in this case the radiation in the wavelength range 0.3 to about 3 μm where the glass of the outer bulb is transparent – will pass through the outer bulb. The thermal radiation of the alumina (for the greater part above 6 μm) will be absorbed almost completely by the glass. The absorbed radiation energy will leave the outer bulb in the form of thermal radiation and – as the outer bulb is normally surrounded by air – in the form of power transferred by conduction and convection to the air.

Discharge Tube Boundary
The radiant power incident on the outer bulb after passing the discharge tube wall is given by

$$P^l_{rad} = P^l_{uv} + P^l_{vis} + P^l_{ir} \tag{2.4a}$$

or

$$P^l_{rad} = t_d(uv)\, P^o_{uv} + t_d(vis)\, P^o_{vis} + t_d(ir)\, P^o_{ir} + t'_d\, f_{rad}\, P_{el} \tag{2.4b}$$

or

$$P^l_{rad} = t_d\, P^o_{rad} + t'_d\, f_{rad}\, P_{el} \tag{2.4c}$$

where
$t_d(\lambda)$ = effective transmission factor of the discharge tube in wavelength range (λ) considered

t_d = effective discharge tube transmission factor for total discharge radiation

t'_d = effective discharge tube transmission factor for electrode radiation

f_{rad} = the fraction of the electrode power emitted as radiation

It may be assumed that the alumina wall of the discharge tube has a high transmission factor ($t_d \approx 0.90$–0.95, depending on material properties, temperature, etc.) in the wavelength range extending roughly from 0.2 to 6 μm (Sec. 8.1.3). Radiation from the discharge or from the electrodes in the far-ultraviolet ($\lambda < 0.2\ \mu$m) and far-infrared ($\lambda > 6\ \mu$m) regions will be absorbed by the alumina wall of the discharge tube. The absorbed radiation as well as the conduction power from the discharge P_{con}^{o} and from the electrodes P_{con}^{el} will leave the wall in the form of thermal ($\lambda > 6\ \mu m$) radiation P_{th}^{l}

$$P_{th}^{l} = P_{con}^{o} + P_{con}^{el} + (1-t_d)\, P_{rad}^{o} + (1-t_d')\, f_{rad}\, P_{el} \tag{2.5}$$

Outer Bulb Boundary

The radiant power is filtered by the outer bulb before leaving the lamp. The radiant power P_{rad} measured outside the lamp (in the wavelength region where the outer bulb is transparent) is given by

$$P_{rad} = P_{uv} + P_{vis} + P_{ir} \tag{2.6a}$$

or

$$P_{rad} = t_b(uv)\, P_{uv}^{l} + t_b(vis)\, P_{vis}^{l} + t_b(ir)\, P_{ir}^{l} \tag{2.6b}$$

or

$$P_{rad} = t_b\, P_{rad}^{l} = t_b\, t_d\, P_{rad}^{o} + t_b'\, t_d'\, f_{rad}\, P_{el} \tag{2.6c}$$

where
- P_{rad} = radiant power
- P_{uv} = ultraviolet radiant power
- P_{vis} = visible radiant power
- P_{ir} = infrared radiant power
- $t_b(\lambda)$ = effective outer bulb transmission factor in wavelength range (λ) considered
- t_b = effective outer bulb transmission factor for incident discharge radiation
- t_b' = effective outer bulb transmission factor for incident electrode radiation

It may be assumed that the glass outer bulb has a high transmission factor ($t_b \approx 0.98$) in the wavelength range extending roughly from 0.3 to 3 μm. The power absorbed by the outer bulb P_b is equal to the difference between the input power P_{in} and the radiant power P_{rad}

$$P_b = P_{in} - P_{rad} \tag{2.7}$$

It follows from Eqs (2.1), (2.2), (2.6) and (2.7) that

$$P_b = P_{con}^{o} + (1 - t_b\, t_d)\, P_{rad}^{o} + P_{el} - t_b'\, t_d'\, f_{rad}\, P_{el} \tag{2.8a}$$

Neglecting the last term in Eq. (2.8a) this equation may be approximated by

$$P_b \approx P_{con}^{o} + (1 - t_b\, t_d)\, P_{rad}^{o} + P_{el} \tag{2.8b}$$

This equation may be written as

$$P_b = P_{c+a} + P_{el} \tag{2.9}$$

where P_{c+a} represents the sum of the conduction losses of the discharge and the absorption losses of the discharge radiation. In its turn, the outer bulb transfers its power to the surroundings in the form of thermal radiation P_{th}, and conduction and convection losses P_{c+c}

$$P_b = P_{th} + P_{c+c} \tag{2.10}$$

The Human Eye
Finally the luminous flux of the lamp is evaluated by the human eye according to its spectral luminous efficiency curve (Sec. 1.2.1). The luminous flux Φ and the luminous efficacy η are calculated from the (absolute) spectral power distribution according to the following equations

$$\Phi = K_m \int\limits_{380}^{780\ nm} P_\lambda\, V(\lambda)\, d\lambda$$

$$= K_m \frac{\int\limits_{380}^{780\ nm} P_\lambda\, V(\lambda)\, d\lambda}{P_{vis}} \qquad P_{vis} = K_m\, V_s\, P_{vis} \tag{2.11}$$

$$\eta = \frac{\Phi}{P_{in}} = K_m\, V_s \frac{P_{vis}}{P_{in}} \tag{2.12}$$

where Φ = luminous flux
K_m = maximum spectral luminous efficacy
P_λ = spectral power distribution
$V(\lambda)$ = spectral luminous efficiency (zero outside the wavelength range 380 to 780 nm)
λ = wavelength
V_s = luminous efficiency of visible radiation
η = luminous efficacy

In Eqs (2.11) and (2.12) the efficiency of the visible radiation for the generation of light is expressed with the aid of the factor V_s, which may be considered as the ratio of the 'light watts' (defined as Φ/K_m) to the 'watts of visible radiation' (P_{vis}).

In this section a qualitative description of the power flow in an HPS lamp has been presented. In the following sections quantitative data will be given.

38

2.2 Electrode Power

Some of the input power – viz. the electrode power – is used to maintain the temperature of the electrodes at the level required for electron emission. This power is dissipated by radiation and conduction. As most of the radiation is infrared, and as most of the conduction power is lost as thermal radiation, this must be considered as a loss for the light production. Therefore this electrode power is also called the electrode losses.

The electrode power can be determined in two ways (see Sec. 9.3.2). It may be derived from a measurement of the electrode fall and the lamp current, their product being assumed to represent the electrode losses. Or, alternatively, the temperature profile along the electrodes may be measured, and the losses due to thermal radiation and heat conduction derived by calculation.

The data found in literature for the electrode losses in 400 W HPS lamps vary considerably (table 2.1), but the following data for the electrode power P_{el} may be considered as representative for typical HPS lamps

$$P_{el} = 8 \text{ W for a } 50 \text{ W HPS lamp}$$
$$P_{el} = 24 \text{ W for a } 400 \text{ W HPS lamp}$$

As roughly 5 to 20 per cent of the input power is lost, the electrode power represents a significant power loss for HPS lamp, especially for the low-wattage lamps.

Table 2.1 Lamp current (I_{la}), electrode fall (V_{el}) and electrode power (P_{el}) in 400 W HPS lamps

	Jack and Koedam (1974)	Denbigh and Wharmby (1976)	Akutsu and Saito (1979)
I_{la} (A)	4.45	4.45	4.0
V_{el} (V)		4 ± 1	9 *
P_{el} (W)	24	17 ± 5	30 *

* Data for electrode fall taken from Ozaki (1971a).

2.3 Radiant Power

The radiant power emitted by the lamp in the wavelength range where the outer bulb is transparent, can be determined from the absolute (spatially integrated) spectral power distribution or from the spatial distribution of the (wavelength integrated) irradiance at a certain distance from the lamp, as measured with a thermopile.

2.3.1 Spectral Measurements

The absolute, spatially integrated spectral power distribution can be measured with a spectroradiometer with the lamp placed in an integrating sphere. The spectroradiometer consists in principle of a monochromator and a detector, for fast measurements an optical multichannel analyzer, i.e. a polychromator with an array of detectors, being used. Measurements of spectral power distribution and luminous flux can also be carried out simultaneously (see figure 2.2). A tungsten halogen incandescent lamp with known luminous flux and spectral power distribution is used for calibration. For the spectral measurements the accuracy is somewhat limited in the infrared region, as the reflectance of the inner surface of the painted sphere and the responsivities of most detectors are relatively low in that wavelength region. But the radiation of the lamp may also be measured directly, i.e. without the integrating sphere shown in figure 2.2. The absolute level of the infrared spectral power distribution is then matched to that in the visible range in the overlap region (de Groot *et al.*, 1975b).

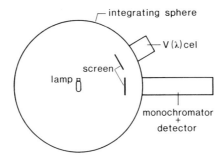

Figure 2.2 Typical set up for the measurement of luminous flux and spectral power distribution. The luminous flux is measured with the aid of a so-called V(λ)-cell (viz. a detector whose spectral responsivity corresponds with the spectral luminous efficiency curve) and the spectral power distribution with a monochromator and detector. The integrating sphere integrates the light and radiation distributions of the lamp spatially.

Typical HPS lamp spectra as measured in the wavelength range 200 to 2500 nm are given in figures 2.3 and 2.4. The following major contributions to the spectrum may be discerned

– The strongly self-reversed sodium D-lines (see also figure 1.9b where a spectrum with higher spectral resolution is given). At high sodium vapour pressures there is a considerable contribution of the sodium D-lines to the spectral power distribution in the wavelength region extending roughly from 540 to 920 nm.

– The non-resonant sodium lines.

40

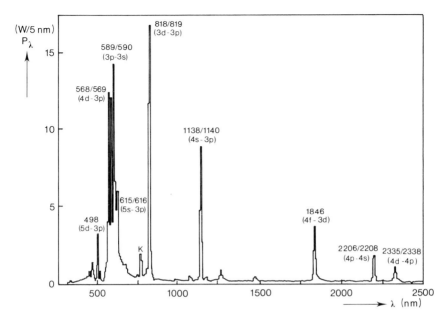

Figure 2.3 Spectral power distribution of a standard 400 W HPS lamp.
The most important sodium lines are identified. The potassium line (K) is due to the presence
of potassium impurities in the amalgam. P_λ is the radiant power in a wavelength interval of
5 nm, λ the wavelength.
Data calculated from this spectrum are:
luminous flux $\Phi = 48000$ lm
luminous efficiency of visible radiation $V_s = 0.57$
chromaticity coordinates $(x,y) = (0.530, 0.415)$
correlated colour temperature $T_c = 2000$ K
general colour rendering index $R_a = 23$

– Lines of impurities (e.g. potassium).
– A continuum in the ultraviolet and blue parts of the spectrum.
– A continuum in the infrared.

A more detailed discussion of the spectrum below 920 nm will be given in the Secs 3.1 and 5.2. It has been shown by Wharmby (1984) that a significant fraction of the infrared continuum is due to thermal radiation from the arc tube, as indicated in figure 2.5. Part of the infrared continuum is also due to thermal radiation from the electrodes (Secs 2.1 and 2.2). The distribution of the radiant power over various lines and continua is given in table 2.2. Discharge radiation above 2500 nm is not included in this table; according to measurements of Wharmby (1984), this radiation makes a relatively small contribution ($\approx 0.7\%$) to the radiant power.

The radiant powers in the ultraviolet range (200–380 nm), in the visible range (380–780 nm) and in the infrared range (780–2500 nm) can be determined

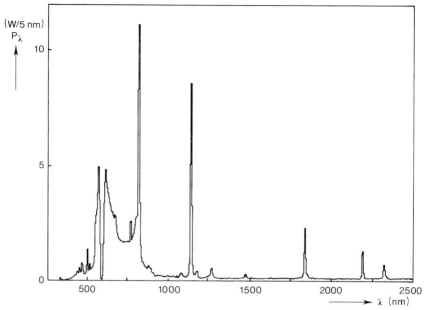

Figure 2.4 *Spectral power distribution of a 'white' 400 W HPS lamp with good colour rendering properties. At such high sodium vapour pressures there is a strong emission in the far wings of the sodium D-lines, extending from 540–920 nm.*
Data calculated from this spectrum are:
luminous flux Φ = 28000 lm
luminous efficiency of visible radiation V_s = 0.34
chromaticity coordinates (x,y) = (0.490, 0.415)
correlated colour temperature T_c = 2350 K
general colour rendering index R_a = 85

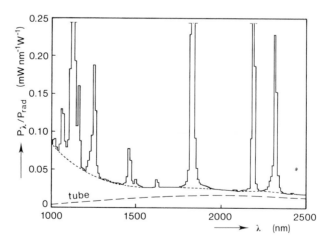

Figure 2.5 *Infrared spectrum of a 400 W HPS lamp showing the thermal radiation from the arc tube (——) and the continuum radiation of the lamp as a whole (--).*
P_λ / P_{rad} is the ratio of the spectral radiant power to the total radiant power. (Wharmby, 1984)

42

from the absolute spectral power distribution by integration. The results for a standard 400 W HPS lamp are compared in table 2.3 with published data. There is good agreement in the values for visible radiation when comparing these data for lamps with about the same luminous flux. The differences in ultraviolet radiation can, in the main, be ascribed to differences in the ultraviolet transmission of the outer bulb (glass or quartz). The differences of 5 to 20 per cent in the values for infrared radiation as derived from spectral measurements (Waymouth, 1977; Wharmby, 1984; present authors), reflect the difficulties encountered when determining the radiation in this part of the spectrum. Part of the differences is due to differences in lamp radiation (indicated by the different luminous flux values) and part is due to measuring uncertainties (estimated to be 5 to 10 per cent). Systematic differences arise

Table 2.2 Division of the radiant power of a standard 400 W HPS lamp over spectral lines and continuum radiation. Data are based on high and low-resolution spectral measurements (figures 1.9b and 2.3)

		Power	
		(W)	(%)
Na 330.2/330.3 nm		0.2	(0.1)
continuum ultraviolet ($\lambda < 380$ nm)		0.7	(0.3)
Na 466.5/466.9 nm		1.5	(0.7)
Na 497.8/498.3 nm		4.0	(1.8)
Na 514.9/515.3 nm		0.7	(0.3)
continuum and other lines ($380 < \lambda < 540$ nm)		7.5	(3.3)
Na 568.3/568.8 nm		11	(4.9)
Na D-lines (540 – 920 nm)		105	(46.9)
Na 615.4/616.1 nm		2.4	(1.1)
K 766.5/769.9 nm		2.5	(1.1)
Na 818.4/819.5 nm		28	(12.5)
Na 107.4/108.3 nm		1	(0.4)
Na 1138/1140 nm		16	(7.1)
K 1169/1177 nm		0.8	(0.4)
Na 1268 nm		2.7	(1.2)
Na 1477 nm		0.7	(0.3)
Na 1846 nm		7.8	(3.5)
Na 2206/2208 nm		3.5	(1.5)
Na 2335/2338 nm		2.3	(1.0)
other lines in infrared		1.7	(0.8)
infrared electrode radiation	($870 < \lambda < 2500$ nm)	8*	(3.6)
infrared continuum	($870 < \lambda < 2500$ nm)	8**	(3.6)
thermal alumina radiation	($1000 < \lambda < 2500$ nm)	8**	(3.6)
integrated radiant power	($200 < \lambda < 2500$ nm)	224	(100%)

* estimated value
** estimated value based on ratio of the thermal alumina radiation to the infrared continuum, as given by Wharmby (1984) (figure 2.5).

from the fact that in some cases the infrared radiation of only the discharge is given (Wharmby, 1984) while in other cases the infrared radiation of the complete lamp (including electrodes and discharge tube radiation) is given (Waymouth, 1977; present authors). This may explain part of the 17 W (17%) difference in infrared radiation as found in the various spectral measurements (see table 2.3).

Table 2.3 Luminous flux and radiant power in the ultraviolet, visible and infrared regions of the spectrum for 400 W HPS lamps

	Jack and Koedam (1974)	Waymouth (1977)	Akutsu and Saito (1979)	Wharmby (1984)	present authors
discharge tube material	PCA	PCA	sapphire	sapphire	PCA
internal diameter (mm)		7.0	8.0	7.4	7.5
electrode spacing (mm)	82	90	78	80	82
sodium mole fraction in amalgam		0.69	0.66	0.69	0.66
xenon pressure at 300 K (kPa)		1.9	4.0	0.7	3.3
lamp current (A)	4.45	5.27	4.0		4.4
lamp voltage (V)	105	84	120		105
luminous flux (lm)	48000	46530	51600	48000[f]	48000
ultraviolet radiation (W)	2[a]			1	1
visible radiation (W)	118[b] (123)	117[d]	140	123	123
infrared radiation (W)	80[c] (75)	95[d]		83	100
radiant power (W)	200	212	247[e]	207	224

a) Quartz outer bulb (in the other cases, glass)
b) The definition of the visible region used here is 380–740 nm, the number between brackets is the corrected value for the present definition of that range, viz. 380–780 nm.
c) The infrared radiation is the difference between the radiant power and the sum of the visible and ultraviolet radiation.
d) Data from G.R. Spears and J.F. Waymouth, GTE Sylvania Lighting Center.
e) The radiant power is the difference between the input power and the sum of thermal non-radiative losses and the electrode losses.
f) Published luminous flux for Thorn 400 W HPS lamps (Cayless and Marsden, 1983).

2.3.2 Thermopile Measurements

The radiant power emitted by the lamp in the wavelength range where the outer bulb is transparent can be determined with a detector whose responsivity is independent of wavelength (in the wavelength region of interest) as is the case for a thermopile. A correction has to be applied for measured (far infrared) thermal radiation from the outer bulb. An integrating sphere cannot be used because the reflectance of its painted inner surface is wavelength dependent. The radiant power follows from the irradiance distribution E_e over a closed surface A around the lamp

$$P_{rad} = \int_A E_e \, dA \tag{2.13}$$

The irradiance distribution is usually measured with the lamp placed at the centre of a spherical surface (figure 2.6). The radiant power then follows from

$$P_{rad} = R_o^2 \int_{\theta=0}^{\pi} \int_{\varphi=0}^{2\pi} E_e(\theta, \varphi) \sin\theta \, d\theta \, d\varphi \tag{2.14}$$

where R_o = sphere radius
        θ = angle with vertical axis
        φ = angle with horizontal axis
        E_e = irradiance

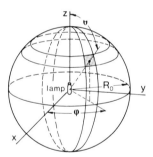

Figure 2.6 *Principle of the measurement of the irradiance distribution on a spherical surface with radius R_0. The detector surface is indicated by the shaded surface, at the angles θ and φ.*

For an HPS lamp operated in a vertical position and with a symmetrical radiation distribution around the vertical axis, Eq. (2.14) may be written as

$$P_{rad} = 2\pi R_o^2 \int_{\theta=0}^{\pi} E_e(\theta) \sin\theta \, d\theta \tag{2.15}$$

For a point source where the irradiance E_e is independent of the direction of observation, Eq. (2.14) can be simplified to

$$P_{rad} = 4\pi R_o^2 E_e \tag{2.16}$$

Considering the HPS lamp as an approximation of a point source, a similar equation can be given for this lamp when introducing a geometric factor g

$$P_{\text{rad}} = g \, R_o^2 \, E_e(\pi/2, 0) \tag{2.17}$$

The geometric factor g is then given by

$$g = \frac{\int\limits_{\theta = 0}^{\pi} \int\limits_{\varphi = 0}^{2\pi} E_e(\theta, \varphi) \sin\theta \, d\theta \, d\varphi}{E_e(\pi/2, 0)} \tag{2.18}$$

Assuming that, for given outer bulb dimensions and lamp construction, the geometric factor is independent of such lamp parameters as current and sodium vapour pressure, the irradiance needs to be measured only in the direction $(\theta = \pi/2, \varphi = 0)$ to determine the radiant power for various lamp parameters, once the geometric factor is known for a certain reference situation.

It is easier to measure the spatial light distribution of a lamp than the spatial radiation distribution. First, because the response time of a silicon cell used to measure the illuminance is much smaller than that of a thermopile; and second, because for all thermopile measurements, a correction has to be applied for the thermal radiation that is emitted by the lamp as well. The geometric factor g for a 400 W HPS lamp with a coated ovoid outer bulb, as deter-

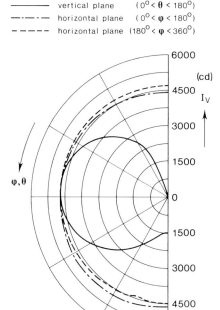

Figure 2.7 *Luminous intensity distribution in various planes for a 400 W HPS lamp with a coated ovoid outer bulb. Burning position is vertical, base up.*

mined from the luminous intensity distribution (figure 2.7), is 10. The same geometric factor can be applied for the radiation measurements as it may be assumed that the light output and the radiation output have approximately the same spatial distribution, provided that the distance between lamp and detector is sufficiently large (at least 1.5 m). This assumption is confirmed by measurements. For the various lamp types the geometric factor varies between 9 and 11, depending on lamp dimensions and outer bulb coating.

As the responsivity of a thermopile is independent of wavelength (in the range 0.15 to 15 μm, for the measurements discussed), most of the thermal radiation of the outer bulb is included in these measurements. The measured radiation is the sum of the radiant power P_{rad} (i.e. discharge and electrode radiation in the wavelength range where the outer bulb is transparent) and the thermal radiation P_{th} of the outer bulb. It is possible to discriminate between these radiation sources because their decay-times differ widely, the decay-time associated with the discharge radiation being much smaller than that for the thermal radiation from the outer bulb, viz. milliseconds as opposed to tens of seconds. A problem is the relatively long response time of the thermopile itself (2.7 s for the measurements discussed). Figure 2.8 shows an example of how the radiant power and the thermal radiation are determined separately from an extrapolation of the thermopile signal, measured as a function of time after switching off the lamp. The 200 W of radiant power as determined with this method by Jack and Koedam (1974) for a 400 W HPS lamp agrees with the results of spectral measurements (207 to 224 W as given in table 2.3) within the accuracy of the measurements (estimated to be 5 to 10 per cent for both methods). It should be noted that with such thermopile measurements it is the radiant power from the discharge that is found and not that from the complete lamp; radiation from electrodes and discharge tube is thereby excluded, as is also the case for the spectral measurements reported

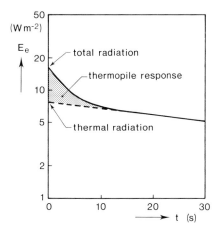

Figure 2.8 Thermopile measurement.
The irradiance E_e (at a distance of 1.5 m from a 400 W HPS lamp) is measured with a thermopile as a function of time t after switching off the lamp. For large values of t the thermopile accurately indicates the thermal radiation (straight line in diagram). The straight line is extrapolated to $t = 0$ (broken line), where it gives the thermal radiation at the moment of lamp switch-off. The discharge radiation is the difference between the total radiation and the thermal radiation at the time $t = 0$.

47

by Wharmby (1984) (Sec. 2.3.1). This may explain, at least partly, why the thermopile measurements of Jack and Koedam (1974), as well as the spectral measurements of Wharmby (1984), give systematically lower values for the radiant power than do the spectral measurements for the complete lamp, as given by Waymouth (1977) and the present authors, see table 2.3.

Such thermopile measurements can, in principle, be used to determine how the power absorbed by the outer bulb is divided between thermal radiation, and conduction and convection losses. The thermal radiation can be determined from an extrapolation of the thermopile signal, measured as a function of time after switching off the lamp (figure 2.8). A geometric factor has to be applied to convert an irradiance value into radiant power [see Eqs (2.17) and (2.18)] and the value for this geometric factor may deviate from the value as found in the foregoing for the measurement of the discharge radiation. The conduction and convection losses are then given as the difference between the input power P_{in} and the sum of radiant power P_{rad} and thermal radiation P_{th} [see Eqs(2.7) and (2.10)].

So far as is known, accurate measurements of the thermal radiation have not been made. An estimation of the ratio of thermal radiation to conduction and convection losses can be made when the outer bulb temperature is known. The thermal radiation can be calculated from the temperature distribution across the outer bulb, using Stefan-Boltzmann's law. The conduction and convection losses also depend on the temperature of the outer bulb. These losses can be calculated with Langmuir's film theory of heat conduction (Dushmann, 1962). For an average outer bulb temperature of 530 K the thermal radiation is much greater than the conduction and convection losses ($P_{th}/P_{c+c} = 2.4$ – see Jack and Koedam, 1974).

2.4 Conduction and Absorption Losses

The sum of conduction and absorption losses P_{c+a} of the column part of the HPS lamp (that is to say excluding the electrode losses) is the difference between input power P_{in} and the sum of radiant power P_{rad} and electrode power P_{el} [see Eqs (2.7) and (2.9)]

$$P_{c+a} = P_{in} - P_{rad} - P_{el} \tag{2.19}$$

Data for the sum of conduction and absorption losses of a (standard) 400 W HPS lamp are given in table 2.4. From these data it follows that this sum amounts to about 40 per cent of the input power of the lamp.

There is yet another method for determining the sum of conduction and absorption losses. This method was employed by Akutsu and Saito (1979) and is based on the assumption that these losses are not significantly higher for the complete lamp than for the discharge tube itself; the radiation absorption

Table 2.4 Comparison of published data for the sum of conduction and absorption losses (P_{c+a}) of 400 W HPS lamps. Relevant data for electrode power P_{el}, radiant power P_{rad} and input power P_{in} are also given

	Jack and Koedam (1974)	Waymouth (1977)	Wharmby (1984)	present authors	Akutsu and Saito (1979)
P_{el}	24 W (6%)	24 W (6.5%)	17 W (4%)	24 W (6%)	30 W (7%)
P_{rad}	200 W (50%)	212 W (53%)	207 W (52%)	224 W (56%)	
P_{c+a}	176 W (44%)	162 W (40.5%)	176 W (44%)	152 W (38%)	123 W*(31%)
P_{in}	400 W (100%)	400 W (100%)	400 W (100%)	400 W (100%)	400 W (100%)

* Direct measurement of the sum of conduction and absortion losses of discharge tube; other data for P_{c+a} are derived from the difference between input power and the sum of electrode and radiant power $P_{c+a} = P_{in} - (P_{el} + P_{rad})$

by the outer bulb is thus neglected. If the conduction power from the electrodes to the discharge tube and the absorption of the electrode radiation may be neglected, then the conduction and absorption losses together are equal to the thermal radiation of the discharge tube [see Eq. (2.5)], and the sum is thus given by the following equation

$$P_{c+a} \approx P'_{th} = 2\pi(R+d)\int_{z=0}^{z=L} \varepsilon(z)\, \sigma\, T_w^4(z)\, dz \qquad (2.20)$$

where P_{c+a} = sum of conduction and absorption losses of the lamp
P'_{th} = thermal radiation of the discharge tube
R = inner radius of the discharge tube
d = wall thickness
z = axial coordinate
L = length of the discharge tube
$\varepsilon(z)$ = (hemispherical) emissivity of discharge tube material
σ = Stefan-Boltzmann constant
$T_W(z)$ = wall temperature as a function of the axial coordinate

Akutsu and Saito (1979) determined the sum of the conduction and absorption losses per unit of discharge tube length by measuring the wall temperature in the homogeneous part, midway between the electrodes ($T_W \approx 1370$ K). Thermal conduction in the axial direction of the discharge tube material is then negligibly small ($\leqslant 1$ per cent). Based on the assumption that the conduction and absorption losses together are independent of the axial coordinate and that the emissivity of the discharge tube material $\varepsilon = 0.25$, they found a value of 123 W for the sum of the conduction and the absorption losses in a 400 W HPS lamp. This value is 20 to 30 per cent lower than other published data (table 2.4), which are based on radiation measurements. The accuracy in the determination of conduction and absorption losses, based on the method of Akutsu and Saito, critically depends on the accuracy with which the PCA emissivity (see also Sec. 8.1.3) and the wall temperature are known.

Applying the two-temperature model of Elenbaas (see Sec. 4.1.1), which gives a linear relation between radiant power and input power, Akutsu and Saito (1979) found a conduction loss at the boundary of the hot core of the arc column of a 400 W HPS lamp of about 50 W ($\approx$ 670 W m^{-1}) for a cold xenon filling pressure of 4 kPa, and a conduction loss of about 110 W ($\approx$ 1400 W m^{-1}) for a neon-argon (99.5%–0.5%) filling with the same pressure. These values are lower limits for the conduction losses of the arc column to the discharge-tube wall as the conduction power is more than doubled in the cooler outer mantle of the discharge (Sec. 4.1.1).

2.5 Power Balances for Various Lamp Types

The results of the measurements described in Secs 2.2 to 2.4 allow the power flow in a standard 400 W HPS lamp to be represented in a diagram, as shown in figure 2.9. The values given are based on the measurements discussed,

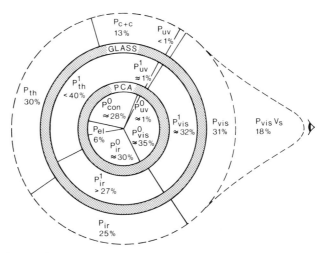

Figure 2.9 The power flow in a standard 400 W HPS lamp following the schematic representation of figure 2.1.
Inside the PCA discharge tube the input power is converted into ultraviolet (P^o_{uv}), visible (P^o_{vis}) and infrared (P^o_{ir}) radiation, electrode power (P_{el}) and conduction power (P^o_{con}). The greater part of the discharge and electrode radiation passes through the alumina wall of the discharge tube (P^l_{uv}, P^l_{vis}, P^l_{ir}), the rest of the power leaving the discharge tube as thermal radiation (P^l_{th}). The greater part of the (near) ultraviolet, visible and infrared radiation passes through the glass outer bulb (P_{uv}, P_{vis}, P_{ir}), the rest of the power leaving the outer bulb as thermal radiation (P_{th}) and by way of conduction and convection (P_{c+c}). The efficiency with which the visible radiation generates light is given by the product of P_{vis} (visible radiant power) and V_s (luminous efficiency of visible radiation).

50

or on estimations. Of the discharge power (94 per cent of input power), some 70 per cent is radiated, so the HPS discharge may be considered as a rather efficient radiation source. Some 35 per cent of the input power is radiated as visible radiation, and as some 10 per cent of this is lost due to absorption in and reflection to discharge tube and outer bulb, eventually 31 per cent being available as useful visible radiation.

It can be shown that the power flow in HPS lamps, differing widely as regards sodium vapour pressure, lamp power and discharge tube dimensions, starting gas and buffer gas, is changed significantly by varying these lamp parameters. The following important quantities are given in figures 2.10 and 2.11

a) visible radiant power P_{vis}, infrared radiant power P_{ir}, electrode power P_{el}, and the sum of the conduction and absorption losses P_{c+a}, all quantities

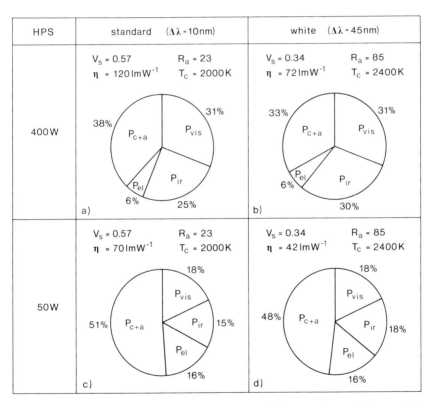

Figure 2.10 Schematic representation of changes in power balance and spectral characteristics when increasing sodium vapour pressure ($\Delta\lambda$-value), or decreasing nominal lamp power (lamp dimensions). For explanation of the symbols see the text.
a) standard 400 W HPS lamp
b) 'white' 400 W HPS lamp
c) standard 50 W HPS lamp
d) 'white' 50 W HPS lamp (extrapolated values)

51

expressed as percentages of the lamp input power. The small amount of ultraviolet radiation is neglected.

b) colour rendering index R_a (as a measure of the colour quality), luminous efficiency of visible radiation V_s and luminous efficacy η.

The result of increasing the sodium vapour pressure (indicated by the wavelength difference $\Delta\lambda$ between the maxima of the self-reversed sodium D-lines) and decreasing the lamp power (dimensions) is shown in figure 2.10. The standard 400 W HPS lamp is taken as a reference. Increasing the sodium vapour pressure to a sufficiently high value results in an HPS lamp emitting white light with good colour rendering properties. The luminous efficacy is lowered mainly because of the lower luminous efficiency of the visible radiation, the amount of visible radiation remaining the same. This is due to the relatively strong emission in the far wings of the sodium D-lines (figure 2.4). There is a slight increase in (near) infrared radiation, while the sum of conduc-

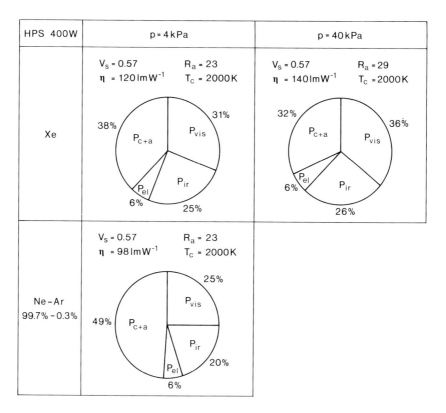

Figure 2.11 Schematic representation of changes in power balance and spectral characteristics of a 400 W HPS lamp when increasing the noble gas pressure or changing the xenon starting gas into a neon-argon mixture. A constant sodium vapour pressure is assumed; p is the noble gas pressure at room temperature.

tion and absorption losses is lowered. Changing the dimensions of the discharge tube (shorter electrode distance, narrower discharge tube) to get a low-wattage HPS lamp, significantly lowers the luminous efficacy. This is because of the relatively higher electrode losses and the relatively higher conduction and absorption losses incurred (Chapters 9 and 4). The relative shape of the spectrum, and thus the light quality and the luminous efficiency of the visible radiation, remain approximately the same. Figure 2.10d shows the extrapolated effects of combined changes in both variables (increased sodium vapour pressure and small discharge tube dimensions). A luminous efficacy of 42 lm W^{-1} may be expected for a white 50 W HPS lamp when other lamp parameters remain constant.

Another important lamp parameter is the noble gas filling, the influence of which is shown in figure 2.11. The standard 400 W HPS lamp is again taken as a reference. Increasing the xenon pressure reduces the combined conduction and absorption losses (Sec. 5.1), while the choice of a neon-argon mixture with a relatively high thermal conductivity, increases the conduction losses. Changes in the spectrum due to the high xenon pressure favour the visible part of the radiation (Sec. 5.2).

2.6 Final Remarks

The purpose of this chapter has been to provide an insight into the power balance and the spectrum of the HPS lamp. It will be clear that the amount of visible radiation and its light quality depend on many lamp parameters. More detailed studies, in combination with theoretical models, will be presented in the following chapters.

Chapter 3

Spectrum and Related Discharge Properties

Important properties of the HPS lamp, such as its chromaticity coordinates, correlated colour temperature, colour rendering, and to a large extent also its luminous efficacy, are governed by the spectral distribution of its radiant power. The spectrum of the high-pressure sodium lamp is very different from that of the low-pressure sodium lamp, with its nearly monochromatic yellow radiation of the sodium D-lines (figure 1.9a). Due to the relatively high sodium density in the HPS lamp, the contours of the sodium D-lines are broadened to such an extent that they cover a large part of the spectrum (figure 1.9b). This broadening is caused by the intrinsic line broadening and by self-absorption.

The integral effect of the local emission and local absorption of radiation in the discharge column is described by the radiative transfer equation (Sec. 3.1.1). In this chapter the relatively simple case of the one-dimensional radiation transport will be treated, while the more complex three-dimensional case will be dealt with in Chapter 4. The calculation of the local emission and absorption of radiation is largely simplified if it may be assumed that the plasma of the HPS discharge is in local thermal equilibrium (LTE) (Sec. 3.1.2). Under this condition the densities in the ground state and the various excited states may be described by one temperature. The spectrum of the HPS lamp is then determined by the plasma temperature distribution, the radius of the discharge tube, the sodium vapour pressure, the type of buffer gas, and its pressure. For the calculation of the spectrum it is necessary to know the plasma composition (Sec. 3.1.3). Also some basic data must be available about the properties of the sodium atom and the sodium molecule and about the broadening of the sodium lines (Sec. 3.1.4). The accuracy of the calculated spectrum can be checked by comparing it with the measured spectrum (Sec. 3.1.5).

The theoretical description of the spectrum, as determined by the sodium vapour pressure, plasma temperature and wall temperature, can be used to obtain information on these parameters from measured spectra (Sec. 3.2).

In order to simplify the theoretical treatment, we shall confine ourselves in this chapter mainly to HPS discharges without a buffer gas; the influence of such a buffer gas will be discussed in Chapter 5. The noble gas present, in the experimental discharges at low pressure for ignition purposes, has only a minor influence on the spectrum (Chapter 5).

3.1 Spectrum of a Pure Sodium Discharge

3.1.1 The One-Dimensional Radiative Transfer Equation

Consider the spectrum as observed looking along a diameter of a cylindrically symmetrical discharge, perpendicular to the discharge axis (figure 3.1). The observed spectral radiance L_v at frequency v can be calculated from the one-dimensional radiative transfer equation describing the integral effect of the local emission and the local absorption of radiation

$$L_v = \int_{-R}^{R} \frac{\varepsilon_v(r)}{4\pi} \exp\left[-\int_{r}^{R} \kappa(v,r')\,dr'\right] dr \qquad (3.1)$$

where L_v = observed spectral radiance at frequency v (per unit solid angle)

r, r' = radial coordinate

R = (inner) radius of discharge tube

$\varepsilon_v(r)$ = spectral emission coefficient

$\kappa(v,r)$ = absorption coefficient

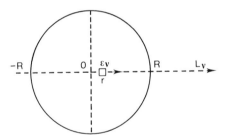

Figure 3.1 *The coordinate system for a cylindrically symmetrical discharge. The local emission of radiation in a certain volume element at radial coordinate r is indicated by ε_v. L_v is the spectral radiance at frequency v in the direction of an observer; R is the radius of the discharge tube.*

The integral of the absorption coefficient along the line of sight is called the optical depth $\tau_0(v)$

$$\tau_0(v) = \int_{-R}^{R} \kappa(v,r)\,dr \qquad (3.2)$$

The emission and absorption coefficients for an LTE plasma are related through Kirchhoff's law

$$\varepsilon_v(r) = \kappa(v,r) \, B_v(T)$$ (3.3)

where $B_v(T)$ is the black body radiation function for frequency v and local temperature $T(r)$, as derived from Planck's law

$$B_v(T) = \frac{2 \, h \, v^3}{c^2} \frac{1}{\exp\left(\dfrac{h v}{k \, T}\right) - 1}$$ (3.4)

where h = Planck's constant
c = speed of light
k = Boltzmann constant

The various spectral quantities given can also be written as a function of wavelength λ instead of frequency v using the relation

$$\lambda = \frac{c}{v}$$ (3.5)

Emission and Absorption Processes

For a high-pressure sodium discharge the following three emission and absorption processes are of major importance, viz. processes involving atomic lines, recombination continuum and molecular bands.

Atomic Lines
Atomic emission lines are due to a transition of an electron in the atom from a high-lying upper level with energy E_u to a lower-lying level with energy E_l (figure 3.2), the frequency v of the line being determined by the relation

$$h v = E_u - E_l$$ (3.6)

The absorption coefficient is given by the relation (Lochte-Holtgreven, 1968)

$$\kappa(v,r) = \frac{1}{4\pi \, \varepsilon_0} \frac{\pi e^2}{m_e c} f \, n_l(r) \, P(v,r)$$ (3.7)

where e = elementary charge
ε_0 = di-electric constant
m_e = electron mass
f = (absorption) oscillator strength
n_l = number density of atoms in state 1
$P(v,r)$ = normalised line profile as a function of frequency v at radial coordinate r

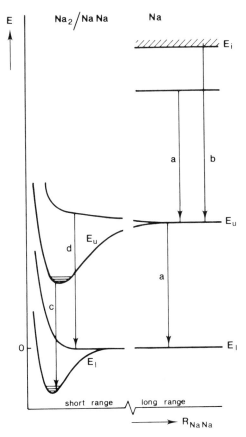

Figure 3.2 Schematic representation of radiation emission processes in sodium atom (Na) and sodium molecule (Na₂) or sodium quasi-molecule (NaNa):(a) atomic lines, (b) recombination continuum, (c) bound-bound bands and (d) free-free bands. The repulsive ground state of the sodium molecule corresponds to a quasi-molecule. The terms short range and long range along the horizontal axis indicate the type of interaction.

R_{NaNa} is the internuclear distance of two sodium atoms, E the atomic or molecular potential energy (index l and u for lower and upper level, respectively), and E_i the ionisation energy.

The line profile $P(v,r)$ describes the frequency distribution of the absorbed and emitted photons as the sodium levels are disturbed by interactions with other sodium atoms and electrons (Sec. 3.1.4). If local thermal equilibrium (LTE) is assumed, the population of the level 1 can be calculated with the aid of the Boltzmann formula from the ground state density, which is related to the sodium vapour pressure and the gas temperature via the equation of state

$$n_1(r) = n_0(r)\frac{g_1}{g_0}\exp\left(\frac{-E_1}{k\,T(r)}\right) \tag{3.8}$$

$$n_0(r) = \frac{g_0}{U_0} n_{Na}(r) = \frac{g_0}{U_0} \frac{p_{Na}}{k\ T(r)} \qquad (3.9)$$

where g_1, g_0 = statistical weight of level 1 and of ground state, respectively
$\quad\quad\ U_0$ = partition function of the atom (approximately equal to g_0 for typical HPS conditions)
$\quad\quad\ n_0$ = number density of atoms in ground state
$\quad\quad\ n_{Na}$ = number density of sodium atoms
$\quad\quad\ p_{Na}$ = (local) sodium vapour pressure

Recombination Continuum

The recombination continuum or free-bound radiation is due to electronic transitions between the ionisation level and the level 1 in the atom (figure 3.2). In fact, an electron (e^-) and an ion (Na^+) are involved in the emission process, which is represented by

$$Na^+ + e^- \rightarrow Na^* + h\nu \qquad (3.10)$$

where Na^* is an excited sodium atom in level 1. The emission coefficient is proportional to the product of ion and electron densities, as given by the Saha equation

$$n_e n_i = 2 n_0 \frac{U_i}{U_0} \frac{(2\ \pi\ m_e k\ T)^{3/2}}{h^3} \exp\left(\frac{-E_i + \Delta E_i}{k\ T}\right) \qquad (3.11)$$

where $\quad n_e$ = (local) electron number density
$\quad\quad\quad n_i$ = (local) ion number density
$\quad\quad\quad U_i$ = partition function of the ion
$\quad\quad\quad E_i$ = ionisation energy
$\quad\quad\quad \Delta E_i$ = lowering of the ionisation energy

The absorption coefficient for the inverse process, which is called photo-ionisation, is given by

$$\kappa(v,r) = \sigma(v)\ n_1(r) \qquad (3.12)$$

where $\sigma(v)$ is the photo-ionisation cross-section for level 1 as a function of the frequency and $n_1(r)$ is the number density of atoms in state 1.

Molecular Bands and Continuum Radiation

Molecular bands and continuum radiation are due to electronic transitions between the energy levels E_u and E_1 in the molecule and can be of the bound-bound, bound-free, free-bound and free-free variety. The free-free continuum joins onto the broadened atomic line radiation (indicated in figure 3.2 by the arrows d and a respectively) and will be called quasi-molecular radiation (radiation from unstable molecules or colliding atoms, see Herzberg (1950) and Gallagher (1975)). In the quasi-static approximation of Hedges *et al.* (1972) and Gallagher (1979) the absorption coefficient for both bound molecules (AB, e.g. Na_2) and quasi-molecules (A + B, e.g. Na + Na) is given by the expression

$$\kappa(v,r) = \frac{1}{4\pi\varepsilon_0} \frac{\pi^2 e^2}{2m_e c} f\, g_1 \; R_{AB}^2 \; \frac{|dR_{AB}|}{|dv|} [A]\,[B]\, \exp\left[\frac{-E_1(R_{AB})}{k\,T(r)}\right] \quad (3.13)$$

where R_{AB} is the internuclear separation of atoms A and B and [A], [B] are the local ground state number densities of the atoms A and B respectively.

In the case of bound molecules the product of the last three factors in Eq. (3.13) is proportional to the molecular density; note that the energy E_1 of the lower level considered (ground state) is the negative dissociation energy.

An important factor in Eq. (3.13) is $\frac{|dR_{AB}|}{|dv|}$. When $\frac{|dv|}{|dR_{AB}|} = 0$ there will be a singularity in the absorption coefficient. This can be understood as follows. When the upper and the lower potential energy curves are parallel to each other (nearly) the same frequency or wavelength will be obtained for transitions at various values of the internuclear distance (figure 3.3). As we may assume that there is a random distribution of equally probable internuclear distances, a maximum in the absorption or emission coefficient must be expected at that frequency (wavelength) where upper and lower potential ener-

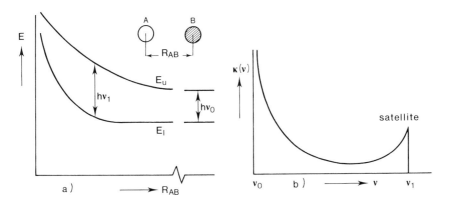

Figure 3.3 Schematic representation of the occurrence of a satellite in the spectrum.
The energies of the upper level (E_u) and the lower level (E_l) are given in (a) as functions of the internuclear separation of the atoms A and B. The resultant absorption coefficient $\kappa(v)$ is given in (b). Apart from the line centre with frequency v_0, there is a maximum in the absorption coefficient in the extreme wing of the line for the satellite frequency v_1, corresponding to that transition where upper and lower potential energy curves are parallel to each other.

gy curves are parallel to each other. Such an extremum in the molecular emission is called a satellite. The singularity at the satellite frequency in the classical quasi-static theory can be avoided by using the modified quasi-static theory of Szudy and Baylis (1975), which decribes the shape of the satellite.

3.1.2 Local Thermal Equilibrium

The term high-pressure discharge indicates not only the order of magnitude of the vapour pressure ($\geqslant 10^4$ Pa), but also that the plasma is approximately in a state of local thermal equilibrium (LTE). This means, for instance, that the gas temperature is approximately equal to the electron temperature. In this sense, the high-pressure sodium discharge is quite different from the low-pressure sodium discharge, where the gas temperature is much lower than the electron temperature (Sec. 1.2.1). For an LTE plasma the various local energy distribution functions, with the exception of that of the radiation energy, are given by equilibrium relationships (Boltzmann distributions) with one temperature as the characteristic parameter. The excited state densities, for instance, are described by the Boltzmann formula (Eq. 3.8), and the electron density by the Saha equation (Eq. 3.11). The characteristic temperature is called the local plasma temperature.

Different estimations of the correctness of the LTE-approximation can be made with the formulae summarised by Richter (see Lochte-Holtgreven, 1968).

For an optically thin, collision-dominated plasma, the LTE state would be reached, if the following condition is fulfilled

$$n_e > 10^{18} \, T_e^{1/2} \, \Delta E^3 \tag{3.14}$$

where n_e = electron density (m^{-3})
T_e = electron temperature (K)
ΔE = highest energy gap in atom (2.1 eV for sodium)

For a sodium plasma with an electron temperature of about 4000 K this implies that the electron density should exceed the value of 10^{21} m^{-3} to approximate the LTE situation. For typical HPS discharges with electron densities of about 10^{22} m^{-3} in the central part of the discharge this condition is thus fulfilled, at least in the central hot channel.

The hot plasma gains the energy by ohmic heating: the electrons acquire their energy from the electric field and this energy is transferred to the atoms and ions by elastic and inelastic collisions. Thus the electron temperature will always exceed the gas temperature, and the temperature difference can be estimated from the following equation

$$\frac{T_e - T_g}{T_e} = \frac{m}{m_e}\left(\frac{\lambda_e e E}{3 k T_e}\right)^2 \tag{3.15}$$

where T_g = gas temperature
m = atom mass
λ_e = mean free path of the electrons
E = electric field strength

For typical high-pressure sodium discharge conditions in the hot channel ($T \approx 4000$ K, $\lambda_e \approx 10^{-7}$ to 10^{-6} m, $E \approx 1000$ V m^{-1}) the calculated difference between electron and gas temperature is 2 to 200 K.

From an analysis of the power flow in an HPS discharge from electrons to atoms and ions, due to elastic and coulomb collisions respectively, Akutsu (1975) concluded that the gas temperature is approximately equal to the electron temperature for sodium vapour pressures above 6 kPa.

Furthermore, for an optically thick plasma, which the HPS discharge is, the radiation absorption plays an important role and this also influences the LTE situation, particularly in the outer region of the discharge. This has been shown by Waszink (1973, 1975) with a non-equilibrium calculation. In his three-level model for the sodium atom, the excitation and ionisation by electron impact, the emission of resonance radiation and the reverse processes were taken into account. The results of his calculations are given in figure 3.4. In the central part of the discharge the difference between the electron temperature and the gas temperature is less than 10 K. In the outer part of the discharge the electron temperature is up to 700 K higher than the gas temperature. This difference in temperature is due to the fact that the strong absorption of the resonance radiation causes the 3p-level to become 'over-populated', thus leading to heating of the electrons by inelastic collisions with excited atoms.

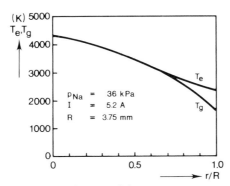

Figure 3.4. *Difference between the electron and the gas temperature.*
Electron temperature T_e and gas temperature T_g were calculated with a non-equilibrium model as functions of the normalised radial coordinate r/R. (Waszink, 1975)

From the foregoing it can be concluded that LTE may be assumed for HPS discharges in calculating quantities that are determined by the conditions in the central hot part of the discharge: quantities such as axis temperature, electric field strength, emission of non-resonant sodium lines and emission in the wings of the sodium D-lines. However, as the plasma is optically thick for the radiation near the centre of the sodium D-lines, this radiation origi-nates from the outer layer of the discharge. The radiation near the centre of the D-lines can therefore only accurately be described by taking into ac-count the deviation from LTE in this outer layer (Sec. 3.1.5).

3.1.3 Plasma Composition

The plasma of a pure sodium discharge consists of sodium atoms (Na) and sodium molecules (Na_2) – either in the ground state or in excited states – sodium ions (Na^+), and electrons (e^-). Assuming that diffusion effects may be neglected, the composition depends only on the total (sodium) pressure and the temperature, which together determine the following equilibria

$$Na_2 \leftrightarrows 2\,Na$$
$$Na \leftrightarrows Na^+ + e^-$$

The total pressure p_{tot} is equal to the sum of the partial pressures of the parti-cles mentioned

$$p_{tot} = p_{Na} + p_{Na_2} + p_{Na}^+ + p_{e^-} \tag{3.16}$$

In a quasi-neutral plasma, the ion and the electron densities are equal. The

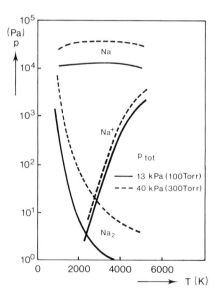

Figure 3.5. The calculated partial pres-sures p of sodium atoms (Na), sodium mol-ecules (Na₂) and sodium ions (Na⁺) as functions of temperature for two values of the total pressure p_{tot}. (de Groot, 1974b)

plasma composition can then be calculated for a given temperature and total pressure from available thermodynamic data (Stull and Prophet, 1971; see also Sec. 5.6). From the calculated partial vapour pressures given in figure 3.5, it can be seen that the sodium molecule density sharply decreases while the ion or electron density sharply rises with increasing temperature. At temperatures around 1000 K (i.e. near the coldest spot of the discharge volume) a significant part (15–20%) of the saturated sodium vapour is in the form of bound molecules (Na_2) (Lapp and Harris, 1966). For discharge temperatures between 1500 and 4000 K the Na vapour pressure is almost equal to the total vapour pressure. At temperatures above 4000 K a significant fraction ($> 1\%$) of the sodium atoms is ionised.

3.1.4 Basic Data Concerning Sodium Atom and Molecule

Sodium Atom

For the calculation of the atomic spectrum of sodium it is necessary to know the energy levels of this atom (figure 3.6) and the oscillator strengths of the lines considered (table 3.1). The broadening of these lines will be discussed in more detail below. The photo-ionisation cross-sections for the sodium 3p and 4s levels are taken from Moskvin (1963) and Rothe (1969). The ionisation energy of the sodium atom is 5.138 eV and the lowering of the ionisation energy can be approximated by

$$\Delta E_i = 6.96 \ 10^{-9} \ n_e^{1/3} \tag{3.17}$$

where the lowering of the ionisation energy is given in eV and the electron density n_e in m^{-3} (Lochte-Holtgreven, 1968).

Broadening Mechanisms of Sodium Lines

The spectral lines as emitted from an elementary volume of a plasma – thus excluding the influence of self-absorption – have a finite line width. This is partly due to the natural line width and the Doppler broadening (caused by the thermal motion of the excited atoms); but the principal cause is that interactions take place between the sodium atoms and their neighbouring particles (perturbers). Due to these interactions a given energy level is subject to a shift ΔE, which for a number of cases can be related to the distance R_{AB} between excited atom A and pertuber B according to the formula

$$\Delta E = \frac{C_p}{R_{AB}^p} \tag{3.18}$$

where C_p is a constant and p is a parameter characterising the type of interaction. The value of p is (Lochte-Holtgreven, Chapter 2, 1968):

Table 3.1 Data for a number of important Na transitions as used for the calculation of the spectrum. The letters l and u denote lower and upper levels respectively; g is the statistical weight, E the energy of the level considered, λ the wavelength, A the transition probability and f the oscillator strength. Data are taken from Wiese et al. (1969)

λ (nm)	Transition	g_l	g_u	E_l (eV)	E_u (eV)	A (10^8 s^{-1})	f
427.36	$3^2P_{1/2} - 10^2D_{3/2}$	2	4	2.10	5.00	0.00391	0.00214
427.68	$3^2P_{3/2} - 10^2D_{5/2}$	4	6	2.10	5.00	0.00469	0.00193
432.14	$3^2P_{1/2} - 9^2D_{3/2}$	2	4	2.10	4.97	0.0055	0.00307
432.46	$3^2P_{3/2} - 9^2D_{5/2}$	4	6	2.10	4.97	0.0066	0.00276
439.00	$3^2P_{1/2} - 8^2D_{3/2}$	2	4	2.10	4.92	0.0081	0.00466
439.33	$3^2P_{3/2} - 8^2D_{5/2}$	4	6	2.10	4.92	0.0097	0.00419
449.42	$3^2P_{1/2} - 7^2D_{3/2}$	2	4	2.10	4.86	0.0126	0.0076
449.77	$3^2P_{3/2} - 7^2D_{5/2}$	4	6	2.10	4.86	0.0151	0.0069
454.16	$3^2P_{1/2} - 8^2S_{1/2}$	2	2	2.10	4.83	0.00359	0.00111
454.52	$3^2P_{3/2} - 8^2S_{1/2}$	4	2	2.10	4.83	0.0072	0.00111
466.48	$3^2P_{1/2} - 6^2D_{3/2}$	2	4	2.10	4.76	0.0214	0.0140
466.86	$3^2P_{3/2} - 6^2D_{5/2}$	4	6	2.10	4.76	0.0257	0.0126
474.79	$3^2P_{1/2} - 7^2S_{1/2}$	2	2	2.10	4.71	0.0059	0.00201
475.18	$3^2P_{3/2} - 7^2S_{1/2}$	4	2	2.10	4.71	0.0119	0.00201
497.85	$3^2P_{1/2} - 5^2D_{3/2}$	2	4	2.10	4.59	0.0418	0.0311
498.28	$3^2P_{3/2} - 5^2D_{5/2}$	4	6	2.10	4.59	0.050	0.0280
514.88	$3^2P_{1/2} - 6^2S_{1/2}$	2	2	2.10	4.51	0.0110	0.00437
515.34	$3^2P_{3/2} - 6^2S_{1/2}$	4	2	2.10	4.51	0.0220	0.00437
568.26	$3^2P_{1/2} - 4^2D_{3/2}$	2	4	2.10	4.28	0.109	0.106
568.82	$3^2P_{3/2} - 4^2D_{5/2}$	4	6	2.10	4.28	0.131	0.095
589.00	$3^2S_{1/2} - 3^2P_{3/2}$	2	4	0.00	2.10	0.630	0.655
589.59	$3^2S_{1/2} - 3^2P_{1/2}$	2	2	0.00	2.10	0.628	0.327
615.42	$3^2P_{1/2} - 5^2S_{1/2}$	2	2	2.10	4.12	0.0241	0.0137
616.07	$3^2P_{3/2} - 5^2S_{1/2}$	4	2	2.10	4.12	0.0482	0.0137
818.33	$3^2P_{1/2} - 3^2D_{3/2}$	2	4	2.10	3.62	0.413	0.830
819.48	$3^2P_{3/2} - 3^2D_{5/2}$	4	6	2.10	3.62	0.495	0.750
1138.1	$3^2P_{1/2} - 4^2S_{1/2}$	2	2	2.10	3.19	0.084	0.163
1140.4	$3^2P_{3/2} - 4^2S_{1/2}$	4	2	2.10	3.19	0.167	0.163
1846.5	$3^2D - 4^2F$	10	14	3.62	4.29	0.140	1.00
2205.6	$4^2S_{1/2} - 4^2P_{3/2}$	2	4	3.19	3.75	0.062	0.90
2208.3	$4^2S_{1/2} - 4^2P_{1/2}$	2	2	3.19	3.75	0.062	0.45
2334.8	$4^2P_{1/2} - 4^2D_{3/2}$	2	4	3.75	4.28	0.067	0.82
2337.9	$4^2P_{3/2} - 4^2D_{5/2}$	4	6	3.75	4.28	0.056	0.91

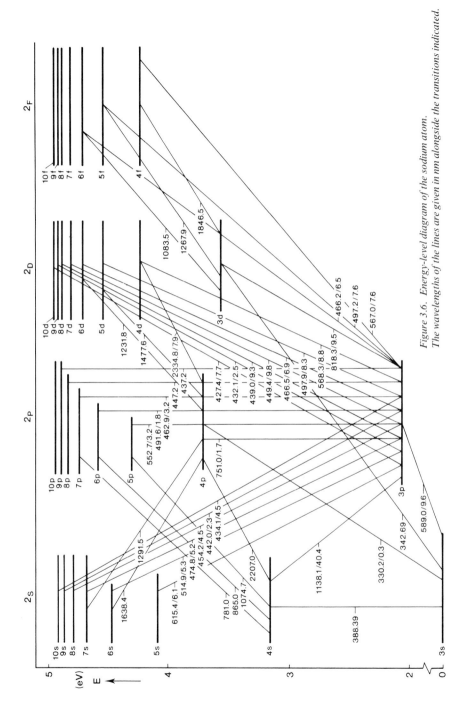

Figure 3.6. Energy-level diagram of the sodium atom. The wavelengths of the lines are given in nm alongside the transitions indicated.

65

p = 2 or 4 for the (linear or quadratic) Stark effect due to interaction of excited sodium atoms with electrons and ions,

p = 3 for resonance interaction between excited sodium atoms and sodium atoms in the ground state,

p = 6 for van der Waals interaction between sodium atoms and all other neutral particles.

The resonance broadening is proportional to the oscillator strength f of the line considered and to the number density n_0 of atoms in the ground state. The full half width $\Delta v_{1/2}$ is given by the relation

$$\Delta v_{1/2} = C_r n_0 \approx \frac{1}{4\pi\varepsilon_0} \frac{e^2}{2\pi m_e v_0} f n_0 \tag{3.19}$$

where C_r is a constant. The various values that are reported in literature for the resonance broadening constant C_r are summarised in table 3.2 for the sodium D_1 and D_2 lines separately as well as a weighted averaged value for both D-lines together. The C_r values for the relevant calculations presented in this book, following de Groot (1972), are based on the data of Foley (1946).

Table 3.2 Comparison of published values for the resonance-broadening constant C_r (see Eq. 3.19) for the Na D_1 and D_2 lines (589.6 nm and 589.0 nm) as well as weighted averaged value for both D-lines together. The remark exp. indicates that the C_r value is based on experimental work; other values are theoretical

	$C_r (10^{-13} \text{ s}^{-1} \text{ m}^3)$			
	Na-D_1	Na-D_2	Na $D_1 + D_2$*	remark
Houston (1938)	0.49	0.49	0.49	
Watanabe (1941)		0.80		exp.
Foley (1946)	0.71	1.01	0.91	
Ali and Griem (1965)	0.50	0.71	0.64	
Reck et al. (1965)	0.55	0.84	0.74	
Carrington et al. (1973)	0.47	0.77	0.67	
Niemax and Pichler (1975)	0.92	1.10	1.04	exp.
Movre and Pichler (1977, 1980)	0.54	0.83	0.73	
Popov and Ruzov (1980)	0.61			
Huennekens and Gallagher (1983)	0.49	0.75	0.66	exp.

$$* C_r (D_1 + D_2) = \frac{f_1 C_r(D_1) + f_2 C_r(D_2)}{f_1 + f_2}$$

In view of the more recently published data, the accuracy of these C_r values is estimated to be between 20 and 30 per cent. The resulting line profile $P(v)$ for resonance broadening can be approximately described by a Lorentz profile (Lochte-Holtgreven, 1968; Movre and Pichler, 1980; Huennekens and Gallagher, 1983)

$$P(v) = \frac{\Delta v_{1/2}}{2\pi} \frac{1}{(v-v_0)^2 + \left(\frac{\Delta v_{1/2}}{2}\right)^2}$$ (3.20)

This approximation is valid for the line core (described by the so-called impact approximation) and the line wings (described by the so-called quasi-static approximation). However, the validity in the far line wings is limited to the wavelength region that corresponds to long-range dipole-dipole Na-Na interaction. At shorter ranges the exchange contribution to the Na-Na interaction leads to a deviation from the Lorentz profile in the very far wings of the D-lines (Niemax and Pichler, 1975; Zemke *et al.*, 1981). In fact there is a continuous transition, as indicated in figure 3.2, from the atomic radiation to the molecular radiation, which will be discussed in the next section.

The Stark effect due to the interaction between sodium atoms and electrons also leads to a Lorentz profile, which may be shifted with respect to the undisturbed line frequency. The broadening half-width and the shift are approximately linearly dependent on the electron density (Griem, 1964; Benett and Griem, 1971; Dimitrijević and Sahal-Bréchot, 1985). The high-lying Na d-levels in particular are strongly influenced by the Stark broadening and shift. In table 3.3 the various broadening mechanisms are compared for a few sodium lines under typical HPS discharge conditions. The resonance interaction

Table 3.3 *Full half-width (FWHM) or shift of the Na 589.0 nm, 616.1 nm and 568.8 nm lines as calculated for various broadening mechanisms under typical high-pressure sodium discharge conditions:* $p_{Na} = 10\,kPa$
a) $T = 4000\,K$ ($n_{Na} = 1.67\,10^{23}\,m^{-3}$, $n_e = 7.0\,10^{21}\,m^{-3}$)
b) $T = 2000\,K$ ($n_{Na} = 3.62\,10^{23}\,m^{-3}$, $n_e = 3.0\,10^{18}\,m^{-3}$)

Broadening mechanism	FWHM/shift (nm)					
	589.0 nm $3^2S_{1/2}-3^2P_{3/2}$		616.1 nm $3^2P_{3/2}-5^2S_{1/2}$		568.8 nm $3^2P_{3/2}-4^2D_{5/2}$	
	4000 K	2000 K	4000 K	2000 K	4000 K	2000 K
Natural line width	0.00001	0.00001	0.00001	0.00001	0.00001	0.00001
Doppler broadening	0.0056	0.0039	0.0058	0.0041	0.0054	0.0038
Resonance broadening *	0.021	0.042	0.023	0.046	0.019	0.039
Stark broadening **	0.002	0.000	0.038	0.000	0.203	0.000
Stark shift **	0.001	0.000	0.021	0.000	0.070	0.000

* Based on data of Foley (1946).
** Based on data of Griem (1964) and Benett and Griem (1971).
p_{Na} is the sodium vapour pressure, T the plasma temperature, n_{Na} the sodium atom number density and n_e the electron number density.

and the Stark effect are the most important broadening effects for the sodium lines considered in the case of a "pure" sodium discharge.

Sodium Molecule

The sodium molecular absorption spectrum can be calculated from Eq. (3.13) when the energy levels and the oscillator strengths of the relevant transitions are known (Woerdman and de Groot, 1981, 1982a). The lowest-lying potential energy curves (see figure 3.7) have been taken from Konowalow *et al.* (1980), Kusch and Hessel (1978), Lyyra and Bunker (1979), and Zemke *et al.* (1981); the higher-lying levels (see e.g. Jeung (1983) have not yet been taken into account. For the oscillator strengths the values of the correlated atomic transitions may be taken.

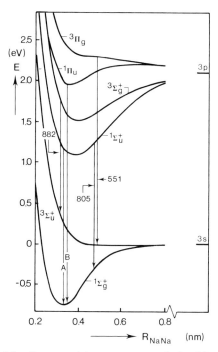

Figure 3.7. Lowest-lying Na_2 potential energy curves, calculated by Konowalow et al. (1980), in dependence of the internuclear separation of the sodium atoms. Arrow A indicates the Na_2 red band system $(A^1\Sigma_u^+ - X^1\Sigma_g^+)$, B the green Na_2 band system $(B^1\Pi_u - X^1\Sigma_g^+)$, 805 the 805 nm satellite of the red band system, 551 the 551.5 nm satellite of the $^3\Pi_g - ^3\Sigma_u^+$ triplet system and 882 the 882 nm satellite of the $^3\Sigma_g^+ - ^3\Sigma_u^+$ triplet system.

Calculated molecular absorption spectra are shown in figure 3.8 for various transitions and for temperatures of 4000 K and 2000 K. To obtain the complete spectrum the contributions of the separate transitions should be added.

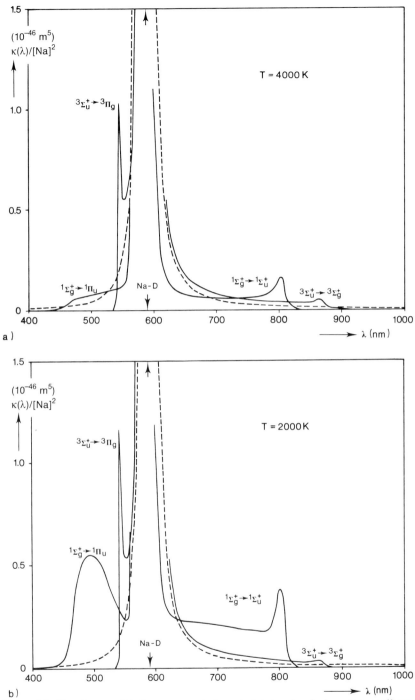

Figure 3.8. The drawn solid lines represent the calculated reduced absorption coefficient $\kappa(\lambda)/[Na]^2$ as a function of wavelength for the Na_2 transitions indicated, for temperatures of (a) 4000 K and (b) 2000 K (Schlejen and Woerdman, 1985). The dashed curves represent $\kappa(\lambda)/[Na]^2$ as calculated for the resonance broadened Na D-lines.

69

The contributions from the bound (Na_2) molecules and from the (NaNa) quasi-molecules show a quite different temperature dependence. The reduced absorption coefficient (absorption coefficient $\kappa(\lambda)$ divided by the sodium atom ground state density squared $[Na]^2$) for the singlet bands decreases with increasing temperature as the bound molecules are strongly dissociated at the high temperatures (figure 3.5). The reduced absorption coefficient for the $^3\Sigma_u^+ - ^3\Pi_g$ triplet transition is hardly temperature dependent, as the energy of the repulsive $^3\Sigma_u^+$ state is practically equal to that of the sodium atom ground state in the relevant range of internuclear distances. The different temperature dependence of the reduced absorption coefficient for the bound molecules and for the quasi-molecules explains the difference in absorption spectra as measured for the same integral sodium densities at high and low temperatures in an arc discharge and in an oven, respectively (de Groot and van Rooyen, 1975). At the high discharge temperatures there is some absorption from the non-resonance lines, but the absorption spectrum in the visible wavelength region is dominated by the D-lines of the sodium atoms and the triplet absorption spectrum of the quasi-molecules. At the low oven temperatures (saturated sodium vapour at temperatures around 900 K), the absorption spectrum, apart from the sodium D-lines, is dominated by the green Na_2 ($X^1\Sigma_g^+ - B^1\Pi_u$) singlet band system (peak near 490 nm) and by the red Na_2 ($X^1\Sigma_g^+ - A^1\Sigma_u^+$) singlet band system (peak near 650 nm at such low temperatures) of the bound molecules.

In the calculated spectra satellites are visible at 545 nm, 803 nm and 862 nm, corresponding with extrema in the potential energy differences for the $^3\Sigma_u^+ - ^3\Pi_g$, $^1\Sigma_g^+ - ^1\Sigma_u^+$ and $^3\Sigma_u^+ - ^3\Sigma_g^+$ transitions, respectively (Veža et al., 1980; Woerdman and de Groot, 1981, 1982a; Konowalow et al., 1983; Huennekens et al., 1984). These calculated satellites correspond with the experimentally observed bands near 551 nm, 805 nm and 882 nm (Sec. 3.1.5).

Figure 3.8 also gives the reduced absorption coefficient for the resonance broadening of the sodium D-lines. From the comparison with the molecular absorption spectrum, it is clear that the validity of the resonance broadening approximation is valid only for wavelengths that are less than 20 to 30 nm away from the centre of the sodium D-lines. For wavelength values farther away from the centre of the D-lines we are in fact in the transition region between atomic and (quasi-)molecular emission.

3.1.5 Calculated and Measured Spectra

According to the data presented in the preceding section, the half-width values of the resonance-broadened sodium D-lines are less than 0.1 nm for typical HPS discharge conditions. So it is clear that line broadening alone cannot explain the broad contour of the D-lines as observed with HPS lamps. This extra broadening is due to the process of self-absorption in the plasma, and can be described quantitatively by the solution of the radiative transfer equation.

The first calculations for the sodium D-lines were published by Cayless (1965). Similar calculations were carried out by Hoyaux (1968), Teh-Sen Jen *et al.* (1969), Lowke (1969), Ozaki (1971b), de Groot (1972), Hassan and Bauer (1974), Waszink (1975) and Waymouth (1977) for various discharge conditions. These calculations were later extended to include the strongest non-resonant sodium lines in the visible and near-infrared wavelength regions and recombination radiation (de Groot, 1974; de Groot and van Vliet, 1975; van Vliet and de Groot, 1983; Dakin and Rautenberg, 1984). Approximations for the molecular radiation (with limited degree of sophistication) were included in only a few publications (de Groot and van Vliet, 1975; Denbigh *et al.*, 1985).

In this section we shall show how the spectrum depends on plasma temperature distribution, sodium vapour pressure and discharge-tube diameter. For simplicity this discussion will be limited to the Na atomic spectrum (so excluding the molecular contribution), as calculated from Eq. (3.1) for a line of sight along a diameter of the discharge tube (figure 3.1). Similarly, the spectrum can be calculated for a line of sight at a distance y from the arc axis and perpendicular to this axis.

Homogeneous Plasma

To begin with, let us consider two homogeneous plasmas at temperatures of 4000 K and 1500 K respectively. From the calculated spectra (figure 3.9, curves a and b) we see that the spectral radiance within a few nm around the centre of the D-lines is equal to the spectral radiance of a black body radiator at the same temperature as the plasma considered. This can also be shown analytically. For a homogeneous plasma, with emission and absorption coefficients independent of the radial coordinate, Eq. (3.1) can easily be integrated with the aid of Eqs (3.2) – (3.5) to give the relation

$$L_\lambda = B_\lambda(T) \ [1 - \exp(-\tau_0(\lambda))] \tag{3.21}$$

where L_λ is the spectral radiance. To make a direct comparison possible with measured quantities, the spectral radiance is given in this section in terms of wavelength units instead of the frequency units, given in previous sections $(L_\lambda = \frac{c}{\lambda^2} L_\nu)$.

At wavelengths where the plasma is optically thick $(\tau_0(\lambda) \gg 1)$, the observed spectral radiance is equal to the black body radiation B_λ. Far in the wings of the lines, the spectral radiance will approach zero as absorption coefficient (and emission coefficient) become zero $(\tau_0(\lambda) \to 0)$.

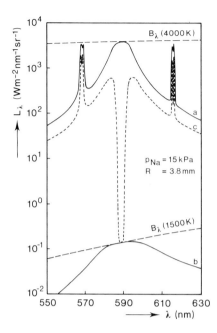

Figure 3.9. Calculated Na atomic spectra for homogeneous sodium plasmas (curve a for T = 4000 K, curve b for T = 1500 K) and for a non-homogeneous plasma (dotted curve c, for a parabolic temperature distribution with axis temperature T_A = 4000 K and wall temperature T_W = 1500 K). The dashed curves give the black body radiation functions $B_\lambda(T)$ for the temperatures T = 4000 K and T = 1500 K.

Inhomogeneous Plasma

For an inhomogeneous plasma, where the temperature in the outer region is lower than that in the central part, the observed contour of the D-lines becomes self-reversed. This is illustrated in figure 3.10 for an imaginary discharge consisting of a 'glowing wire' of hot sodium vapour at the axis, surrounded by cold sodium vapour. The emission near the centre of the sodium D-lines is completely absorbed by the cold sodium vapour. Only the emission in the very far wings of the sodium D-lines can leave the 'discharge' tube. The observed contour is three orders of magnitude broader than the emitted line profile. While the emitted Lorentz profile (without self-absorption), in the case considered, has a half-width value of only 0.025 nm, the wavelength separation between the self-reversal maxima of the observed contour amounts to 11.5 nm.

Consider now a more realistic temperature distribution described by the relation

$$T(r) = T_A - (T_A - T_W) \left(\frac{r}{R}\right)^{n_T} \tag{3.22}$$

where T_A = axis temperature

T_W = wall temperature

n_T = parameter describing the relative temperature profile

72

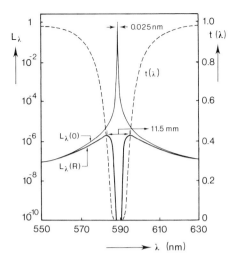

Figure 3.10. Illustration of self-reversal of the sodium D-lines.

In this case the D_1 and D_2 lines are considered as a single line with the line centre at 589.2 nm. L_λ (0) gives the normalised spectral radiance in the arc centre, which is assumed being proportional to the emission coefficient in the arc centre, $t(\lambda)$ gives the transmission for the plasma between arc centre and wall and L_λ (R) the observed spectral radiance at the arc tube radius R. $T(r) = 2500$ K except at the axis, where $T(0) = 4000$ K, $p_{Na} = 15$ kPa, $R = 3.8$ mm.

The calculated spectrum for a parabolic temperature distribution ($n_T = 2$ in Eq. 3.22) is given as curve c in figure 3.9. The spectral radiance is relatively low near the centre of the D-lines (where the optical depth of the discharge is great), as this radiation comes effectively from the cool outer mantle of the discharge. The observed radiance is then equal to the radiance of a black body with a temperature equal to the wall temperature. Farther away from the line centre the spectral radiance increases, because the observed radiation is generated nearer to the hot central part. Still farther away from the line centre, very far in the wings, the spectral radiance again decreases because of the low emission coefficient in the hot central part, as in case of a homogeneous plasma.

Now we shall consider the effects of changing the axis temperature, the sodium vapour pressure and the discharge tube radius on the spectrum of the HPS discharge. For some typical cases, these spectra have been calculated for the visible wavelength region with integration intervals of 1 nm (see figures 3.11 and 3.12). From these spectra the following conclusions can be drawn

a) Axis Temperature

An increase in axis temperature causes a relatively strong increase in the observed spectral radiance (figure 3.11). The relative contour of the sodium D-lines remains nearly the same. The relative increase in radiation is stronger for radiation originating from higher lying levels. This causes an increase in the intensity of the non-resonance lines relative to that of the sodium D-lines. In particular the blue sodium lines originating from the high-lying d-levels and the recombination radiation in the blue-violet region of the spectrum are more clearly visible at the higher axis temperatures. These changes

73

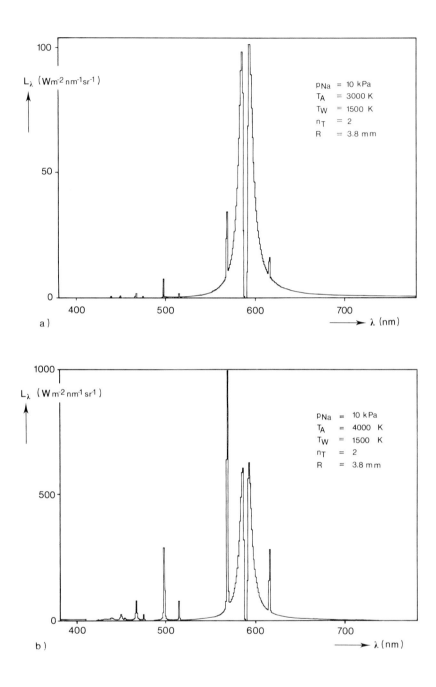

Figure 3.11. Influence of the axis temperature on the Na atomic spectrum in the visible wavelength region. The calculated spectral radiance L_λ is given as a function of wavelength (1 nm integration interval) for two values of the axis temperature: a) $T_A = 3000\ K$, b) $T_A = 4000\ K$.

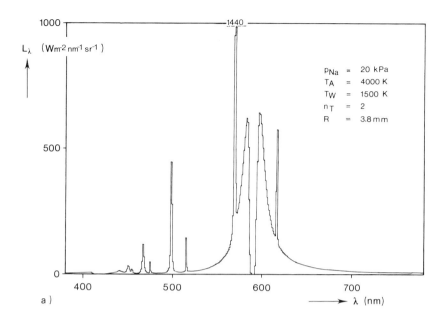

a)

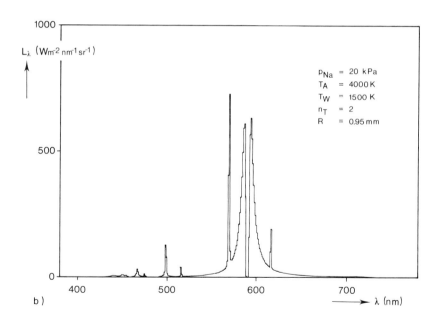

b)

Figure 3.12. Influence of discharge-tube radius on the Na atomic spectrum in the visible wavelength region. The calculated spectral radiance L_λ is given as a function of wavelength for two values of the discharge-tube radius: a) $R = 3.8$ mm, b) $R = 0.95$ mm. The sodium vapour pressure p_{Na} is a factor of 2 higher than in figure 3.11

75

in the visible spectrum have a clear influence on the colour of the light emitted (figure 3.13a). At higher axis temperatures there is a considerable increase in the correlated colour temperature T_c. A higher axis temperature causes only a modest increase in the general colour rendering index. The axis temperature can be influenced significantly in practical HPS lamps by modifying the current waveform as will be discussed in Sec. 7.2.

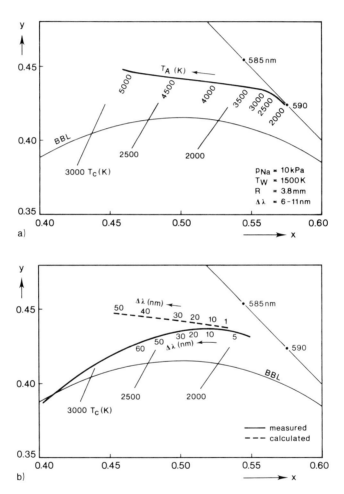

Figure 3.13. Calculated influence of axis temperature and sodium vapour pressure on the colour point of a pure HPS discharge: (a) with varying axis temperature T_A, calculated for a constant sodium vapour pressure, and (b) with varying sodium vapour pressure calculated for a constant temperature distribution $(T_A = 4000\ K,\ T_W = 1500\ K,\ n_T = 2)$ and a discharge-tube radius $R = 3.8$ mm. The self-reversal widths of the sodium D-lines $(\Delta\lambda)$ are indicated in figure (b) along the curves; measured colour points for 'pure' sodium discharges with different $\Delta\lambda$ values have been approximated by the full curve.

b) Sodium Vapour Pressure

An increase in the sodium vapour pressure causes an increase in the self-reversal width of the sodium D-lines (compare figures 3.11b and 3.12a). The spectral radiance in the self-reversal maxima is approximately independent of the sodium vapour pressure as this radiance is determined mainly by the plasma temperature. An increase in the sodium vapour pressure further increases the spectral radiance of the non-resonant sodium lines compared with that of the self-reversal maxima of the sodium D-lines. The combined changes of the broadened D-lines and the stronger radiation in the non-resonance lines with increasing sodium vapour pressure, cause the colour point of the resulting spectrum to shift away from that of monochromatic yellow radiation, giving a higher correlated colour temperature (figure 3.13b). This calculated effect is qualitatively in agreement with experimental results. For a direct comparison between calculations and measurements the wavelength separation $\Delta\lambda$ between the maxima of the self-reversed sodium D-lines is given as parameter in figure 3.13b. Calculated and measured colour points agree reasonably well for $\Delta\lambda$ values between 10 and 30 nm. The considerable differences in colour points at higher $\Delta\lambda$ values are mainly due to the fact that the molecular contribution to the spectrum is neglected in these calculations. These contributions play an important role at the higher $\Delta\lambda$ values, as will be shown below. The differences in colour points at low $\Delta\lambda$ values are probably caused by differences in plasma temperature (for the calculated spectra the axis temperature is taken constant, $T_A = 4000$ K, while for the measured sodium discharges the axis temperature is not constant).

With the increase in self-reversal width of the sodium D-lines at higher sodium vapour pressures, the general colour rendering index R_a increases until

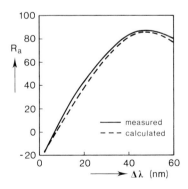

Figure 3.14 *Increase in general colour rendering index R_a with increase in wavelength separation $\Delta\lambda$ between the self-reversal maxima of the sodium D-lines. The calculated values are derived from calculations of the atomic sodium spectrum ($T_A = 4000$ K, $T_W = 1500$ K, $n_T = 2$, $R = 3.8$ mm).*

a maximum of about 85 is reached for a $\Delta\lambda$ value of about 45 nm (figure 3.14). For still higher values of the self-reversal width, the general colour rendering index again decreases. Calculated and measured values for the colour rendering index agree very well, also for high $\Delta\lambda$ values (figure 3.14). The colour rendering index is not so critically dependent on the spectral details as is, for instance, the colour point.

The changes in the spectrum at higher sodium vapour pressures cause the luminous efficiency of the visible radiation to decrease with increasing self-reversal width of the sodium D-lines (figure 3.15). At the higher sodium vapour pressures the radiation is spread more over the blue and red parts of the visible spectrum where the spectral luminous efficiency is relatively low (see also figures 1.9b and 2.4). The calculated luminous efficiency is higher than measured (figure 3.15), as the molecular emission in the far wings of the D-lines is neglected. The changes in spectral distribution of the visible radiation is the main cause of the lower luminous efficacy of HPS lamps at higher sodium vapour pressures (Secs 2.5 and 10.1.3).

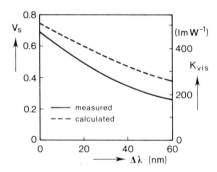

Figure 3.15 Lowering of the luminous efficiency V_s and luminous efficacy K_{vis} of the visible radiation with increase in wavelength separation between the self-reversal maxima of the sodium D-lines. The calculated values are derived from calculations of the atomic sodium spectrum ($T_A = 4000$ K, $T_W = 1500$ K, $n_T = 2$, $R = 3.8$ mm). The measured values are derived from measurements on 'pure' sodium discharges.

From calculations of the spectrum it follows that the wavelength separation $\Delta\lambda$ between the maxima of the self-reversed sodium D-lines depends approximately linearly on the sodium vapour pressure. This can also be shown analytically from Eqs (3.7) and (3.9), provided $\Delta\lambda$ is (much) larger than the wavelength separation between the two D-lines (0.6 nm). In that case the two D-lines may be considered as a single line and the wavelength or frequency difference between the self-reversal maxima and the centre of the D-lines is large as compared to the half-width of the line profile (thus $(v-v_0)^2 \gg (\Delta v_{1/2}/2)^2$ in Eq. 3.20). The optical depth at the self-reversal maxima ($v = v_m$, $\lambda = \lambda_m$) can then be written with the aid of Eqs (3.2), (3.5), (3.7), (3.9), (3.19) and (3.20) as

$$\tau_0(\lambda_m) = \int_{-R}^{R} \kappa(\lambda_m)\ dr = C_1\ \frac{p_{Na}^2}{\Delta\lambda^2}\ R \int_{-1}^{1} \frac{1}{T(x)^2}\ dx \qquad (3.23)$$

where C_1 is a constant and x is the normalised radial coordinate (r/R). At the self-reversal maxima the optical depth has a certain value $(\hat{\tau}_0 \approx 2)$, which is almost independent of the temperature profile, as follows from line contour calculations (see also Bartels, 1950, 1951). This leads to the relation

$$\Delta\lambda = c_1^{1/2}\, p_{Na}\, R^{1/2} \left[\int_{-1}^{1} \frac{1}{T(x)^2}\, dx \right]^{1/2} \tag{3.24}$$

c) Discharge-Tube Radius

The calculated effect of a reduction in discharge tube radius is shown in figure 3.12. The effect on the spectrum is qualitatively comparable to that caused by a reduction in sodium vapour pressure. However, the effect of the discharge-tube radius on the self-reversal width $\Delta\lambda$ of the sodium D-lines is weaker than that of the sodium vapour pressure. From Eq. (3.24) it follows that $\Delta\lambda$ is proportional to the square root of the discharge-tube radius (for a given sodium vapour pressure and temperature distribution), while $\Delta\lambda$ depends linearly on the sodium vapour pressure. A reduction by a factor of 4 in the discharge-tube radius (compare figures 3.12 a and b) thus has the same effect on the contour of the sodium D-lines as a reduction by a factor of 2 in the sodium vapour pressure (compare figures 3.12a and 3.11b). Note that the effect on the intensity of the non-resonance lines is about the same for changes in diameter as for changes in sodium vapour pressure as these lines are approximately equally dependent on the sodium vapour pressure and the discharge-tube radius. A reduction in discharge-tube diameter therefore causes a reduction in the power of the non-resonant sodium lines as compared with that of the sodium D-lines (compare figures 3.11b and 3.12b).

Calculated and Measured Spectra Compared

The accuracy of the calculated spectra can be checked by comparing them with measured spectra. The calculated spectrum must then be corrected for wall transmittance and for multiple wall reflections, taking into account the absorption of the plasma. In a first order approximation this can be done with the formula

$$L_\lambda = L_\lambda\,(\text{calc.})\, \{(\tau_W + \varrho_W\, \tau_W\, \exp[-\tau_0(\lambda)]\}\} \tag{3.25}$$

where $L_\lambda\,(\text{calc.})$ = spectral radiance calculated from the radiative transfer equation

       τ_W = in-line transmittance of the wall

       ϱ_W = reflectance of the wall

       $\tau_0(\lambda)$ = optical depth of the plasma

In the case of a sapphire discharge tube and a glass outer bulb $\tau_w \approx 0.8$ and $\varrho_w \approx 0.2$. Calculated and measured spectra are compared in figure 3.16. The calculated and measured contours of the Na 568.3/568.8 nm (4d – 3p) and the Na 818.3/819.5 nm (3d – 3p) lines are compared in the figures 3.17 and

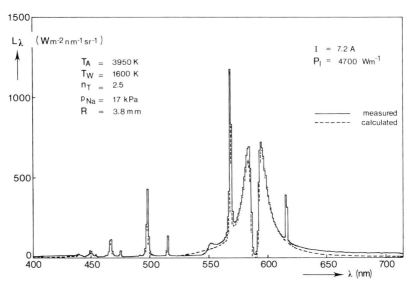

Figure 3.16 *Comparison between the calculated and measured visible spectra of a high-pressure sodium discharge with a sodium vapour pressure of about 17 kPa. The input data for the calculations, such as temperature distribution, sodium vapour pressure p_{Na} and arc-tube radius R, are given in the figure along with the discharge current I and the input power P_l per unit of arc length. The spectra are given with 1 nm integration intervals; the line of sight is along a diameter of the discharge tube. (based on de Groot and van Vliet, 1975)*

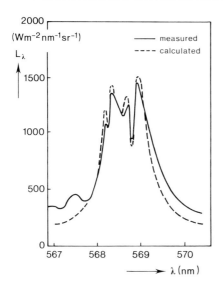

Figure 3.17 *Comparison between the calculated and measured contours of the Na 568.3/568.8 nm lines; the same data apply as in figure 3.16. The lines are self-reversed and they are asymmetric due to the Stark shift. The two humps at 567.0 and 567.6 nm are due to forbidden transitions (4f-3p) in the sodium atom, which are not taken into account in the calculations. (de Groot and van Vliet, 1975)*

80

3.18, respectively. It should be noted that these calculations and measurements are not quite independent, since the data for plasma temperature and sodium vapour pressure used for the calculation of the spectrum were determined from spectral measurements. Considering the uncertainties in the sodium vapour pressure, in the plasma temperature and in the line-broadening data, there is good agreement between the calculated and measured line contours. This is also the case for the calculated and measured total visible spectrum. This agreement supports the validity of the assumption of LTE for the major part of the discharge as the various sodium lines arise from low and high-lying levels, with a large variation in optical depth.

The deviation from LTE in the outer region of the discharge (Sec 3.1.2) influences the contour of the sodium D-lines near their centre, where the measured spectral radiance is determined by the conditions in the outer mantle of the discharge. Therefore, a good description of the spectral radiance near the centre of the D-lines can only be given by non-LTE calculations, taking into account the difference between the electron temperature (describing the pop-

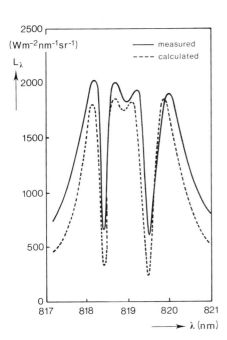

Figure 3.18 Comparison between the calculated and measured contours of the Na 818.3/819.5 nm lines; the same data apply as in figure 3.16. (de Groot and van Vliet, 1975)

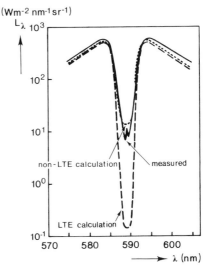

Figure 3.19 Comparison between the calculated and measured contours of the sodium D-lines showing the difference between the LTE and non-LTE calculations. (de Groot et al., 1975a)

The two emission minima at 589.0 and 589.6 nm in the measured contour may be a result of the lower population of the 3p-level within a few free-path lengths from the tube wall.

81

ulation of the 3p-level) and the gas temperature, see figure 3.19. The spectral radiance at the self-reversal maxima of the D-lines and the radiance of the non-resonant sodium lines are hardly influenced by the deviation from LTE in the outer mantle of the discharge as the radiation at these wavelengths is generated mainly in the central, hot part of the discharge.

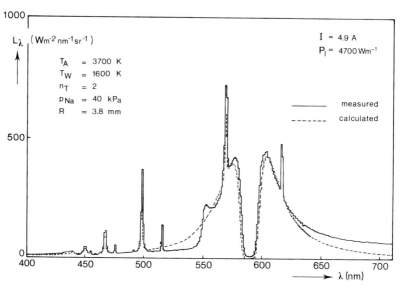

Figure 3.20 Comparison between the calculated and measured spectra of a high-pressure sodium discharge with a sodium vapour pressure of about 40 kPa. (based on de Groot and van Vliet, 1975)

Further there are deviations between the calculated and the measured spectra in the far wings of the sodium D-lines: the measured radiance is higher than calculated in the red wing and also in the blue wing for the band near 551.5 nm. These effects are more pronounced at the higher sodium vapour pressures (figure 3.20). At relatively high sodium vapour pressures there is a significant contribution to the spectrum from the very far wings of the sodium D-lines where (quasi-)molecular emission is dominant. This is shown in figure 3.21 where some spectra are given in the wavelength range 350–950 nm for an experimental high-pressure sodium discharge showing a strong self-reversal of the sodium D-lines. This experimental discharge tube was ovoid-shaped, giving a strong D-line reversal for the section of the discharge tube midway between the electrodes, at a relatively low wall loading. Because of the resulting relatively low axis and wall temperatures, in combination with the relatively large discharge-tube radius ($R = 9$ mm), the molecular contribution is much more pronounced than in standard HPS discharges. For the spectra with a self-reversal of the sodium D-lines as extreme as those shown

82

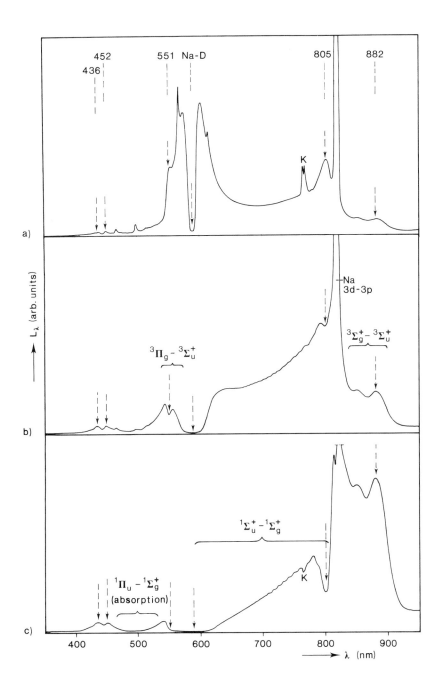

Figure 3.21 Spectra of an experimental high-pressure sodium discharge at increasing input power and sodium vapour pressure, showing the strong molecular contribution at the extreme self-reversal widths of the sodium D-lines. The discharge tube was especially designed to obtain a strong self-reversal at a low wall loading. The various spectral features are indicated in the figure.

in figure 3.21b and c, the self-reversal width ($\Delta\lambda$) is difficult to define. The molecular character of the very far red wing of the D-lines becomes visible by the vibrational structure of the red $^1\Sigma_u^+ - ^1\Sigma_g^+$ band system. The satellites of the $^3\Pi_g - ^3\Sigma_u^+$ and $^1\Sigma_u^+ - ^1\Sigma_g^+$ transitions at 551 nm and 805 nm respectively are self-reversed. A band can be discerned at 882 nm; this is probably the satellite of the $^3\Sigma_g^+ - ^3\Sigma_u^+$ transition, as can be deduced by comparison with the calculated spectra given in figure 3.8 (see also Huennekens *et al.*, 1984). In the blue-violet region of the spectrum two diffuse bands are visible, one at 436 nm and one at 452 nm (Bartels, 1932; de Groot, 1974b). These so-called violet bands recently attracted many researchers because of the possibility for generating laser action (see Pichler *et al.*, 1983, and references cited herein). They probably originate from a transition from the $2^3\Pi_g$ state (with the Na-3d level as the asymptote) to the $^3\Sigma_u^+$ ground state (Pichler *et al.*, 1983; Li and Field, 1983). The radiation in the wavelength range 470 to 510 nm probably is for the greater part absorbed by the bound Na_2 molecules ($^1\Sigma_g^+ - ^1\Pi_u$ transition) in the cool outer mantle of the discharge.

For the extremely strong self-reversal of the sodium D-lines shown in figure 3.21c the visible spectrum in fact consists of the blue-violet continuum of the violet bands, the green continuum of the $^3\Pi_g - ^3\Sigma_u^+$ satellite near 551 nm, and the red band/continuum of the Na_2 $^1\Sigma_u^+ - ^1\Sigma_g^+$ transition. These three continua give a blueish-white light with a correlated colour temperature over 5000 K.

It will be clear that for a realistic description of such spectra the molecular contribution can no longer be neglected. A reasonable description of the visible spectrum in terms of atomic radiation is only possible for self-reversal widths of the sodium D-lines up to about 30 nm.

3.2 Measurement of Spectrum-Related Discharge Properties

3.2.1 Sodium Vapour Pressure

Usually, in HPS lamps, not all of the sodium present is vaporised. Therefore the sodium vapour pressure can be derived from the temperature of the coldest spot in the discharge tube (Sec. 5.6). This temperature may be controlled by immersing the coldest spot in an indium or tin bath of known temperature (figure 3.22). This method has been employed by several authors (Schmidt, 1963, 1965; Ozaki, 1971b, 1971c; de Groot, 1974b; Denbigh, 1974; Waszink and Flinsenberg, 1978; Inouye *et al.*, 1979; Zollweg and Kussmaul, 1983; Wharmby, 1984; Denbigh *et al.*, 1985). This method cannot be used for a complete HPS lamp where the discharge tube is mounted in an outer bulb. It is then preferable to employ the spectroscopic method developed by Teh-Sen Jen *et al.* (1969).

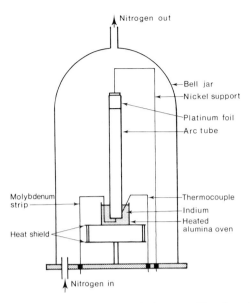

Nitrogen out

Bell jar
Nickel support
Platinum foil
Arc tube

Molybdenum
strip

Heat shield

Thermocouple
Indium
Heated
alumina oven

Nitrogen in

Figure 3.22 Example of the sort of equipment used when controlling the sodium vapour pressure with the aid of an indium bath. (Denbigh, 1974)

Spectroscopically, the sodium vapour pressure p_{Na} is determined from the wavelength separation $\Delta\lambda$ between the maxima of the self-reversed sodium D-lines. An analytical relation between p_{Na} and $\Delta\lambda$ follows from Eq. (3.24)

$$p_{Na} = C_1^{-1/2} \, \Delta\lambda \, R^{-1/2} \left[\int_{-1}^{1} \frac{1}{T(x)^2} \, dx \right]^{-1/2} \tag{3.26}$$

The influence of the plasma temperature distribution on the relation between sodium vapour pressure and $\Delta\lambda$ is rather weak for practical temperature distributions (figure 3.23). As a rule-of-thumb the following relation may be used to determine the sodium vapour pressure

$$p_{Na} \approx \frac{2.7 \pm 0.5}{\sqrt{R}} \Delta\lambda \tag{3.27}$$

where the sodium vapour pressure p_{Na} is given in kPa, the radius R in mm and the self-reversal width $\Delta\lambda$ in nm. This relation is valid for practical HPS discharge conditions ($T_A \approx 4000$ K, $T_W \approx 1500$ K, $n_T \approx 2.5$). The value of the constant C_1 in Eq. (3.26) is proportional to the constant C_r for resonance broadening (table 3.2). In view of the uncertainties in C_r (Sec. 3.1.4) and in the assumed plasma temperature distribution, the inaccuracy in the relation between p_{Na} and $\Delta\lambda$ is estimated to be between 15 and 25 per cent. The analytical relation (3.26) is valid only so long as the self-reversal width is significantly larger than the wavelength separation of the two D-lines ($\Delta\lambda > 0.6$ nm) and so long as resonance broadening is the dominant broadening mechanism at the reversal maxima of the sodium D-lines. As molecular emission makes a significant contribution to this broadening for wavelengths more than 20 to 30 nm away from the line centre, relation (3.26) is no longer valid for $\Delta\lambda$ values larger than 40 to 60 nm. At such high $\Delta\lambda$ values the sodium vapour pressure can be found by fitting calculated spectra (including molecular emission, and with the sodium vapour pressure as the variable) to the measured spectrum.

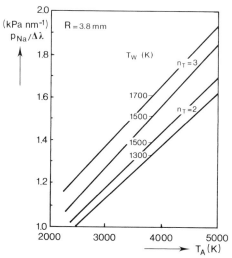

Figure 3.23 *The calculated ratio of the sodium vapour pressure p_{Na} to the wavelength separation $\Delta\lambda$ between the self-reversal maxima of the sodium D-lines as a function of axis temperature T_A for a number of wall temperatures T_W, for a parabolic $(n_T = 2)$ and a cubic $(n_T = 3)$ temperature distribution. The radius of the discharge tube $R = 3.8$ mm. (de Groot, 1974b)*

Calculations show that the self-reversal width of the sodium D-lines is almost independent of the distance between the line of sight and the arc axis. It may be expected, therefore, that relation (3.27) will also be approximately valid for measurements where the spectrum is integrated over the whole cross-section of the discharge or for measurements in an integrating sphere. The self-reversal width is to some extent dependent on the reflectance of the arc tube. Experimentally, slightly larger $\Delta\lambda$ values are found for discharge tubes made of polycrystalline alumina than for sapphire tubes (Zollweg and Kussmaul, 1983). This is ascribed to the back scattering of radiation into the discharge and the consequently greater absorption at wavelengths between the reversal maxima.

To make a direct comparison possible between the results of this spectroscopic method and those of the above-mentioned bath method, an overview is given in table 3.4 of the ratios of sodium vapour pressure $p_{Na}(T_B)$ (as de-

Table 3.4 Overview of the values for $\dfrac{p_{Na}(T_B)}{\Delta\lambda}\sqrt{R}$ as deduced from published metal bath experiments

author	$\dfrac{p_{Na}(T_B)}{\Delta\lambda}\sqrt{R}$ $(\text{kPa nm}^{-1}\,\text{mm}^{1/2})$
Schmidt (1963b)	≈ 2.5
Schmidt (1965)	≈ 1.2
Ozaki (1971b)	≈ 1.8
Ozaki (1971c)	$1.4-2.8$
de Groot (1974b)	≈ 2.4
Zollweg and Kussmaul (1983)	≈ 1.9
Wharmby (1984)	≈ 1.7
Shao-Zhong et al. (1985)	≈ 1.5

$p_{Na}(T_B)$ = sodium vapour pressure as derived from bath temperature

$\Delta\lambda$ = wavelength separation between the maxima of the self-reversed sodium D-lines (published value or derived from published spectrum)

R = (inner) radius of discharge tube

duced from bath temperature T_B) to self-reversal width $\Delta\lambda$ (published value or deduced from published spectrum). This ratio is multiplied by the square root of the radius to make a direct comparison possible with Eq. (3.27). Leaving the extreme values out of consideration the results of table 3.4 can be summarised as follows

$$p_{Na}(T_B) \approx \frac{1.9 \pm 0.5}{\sqrt{R}}\Delta\lambda \qquad (3.28)$$

where $p_{Na}(T_B)$ is given in kPa, R in mm and $\Delta\lambda$ in nm.

Although the numerical values in Eqs (3.27) and (3.28) overlap each other because of the rather large uncertainties, there is strong evidence that the spectroscopic method generally yields higher values for the sodium vapour pressure than does the metal bath method. A possible explanation for such systematic differences is that the metal bath method may underestimate the sodium vapour pressure. This is because, due to heat fluxes, the temperature of the liquid sodium inside the lamp may be higher than the bath temperature (see also Sec. 5.6). Another explanation is that the spectroscopic method overestimates the sodium vapour pressure, the resonance broadening theory underestimating the broadening of the sodium D-lines in the relevant wavelength range because of the molecular effects mentioned.

An accurate comparison between the two methods, preferably in combination with an independent method (e.g. using lamps with known sodium filling, where all the sodium is in the vapour phase), would be desirable.

3.2.2 Plasma Temperature Distribution

Methods of Measurement

Most methods used to measure the plasma temperature in an HPS discharge are spectroscopic methods, based on the measurement of line contours or line intensities. Usually, 'side-on' measurements are performed, viz. the integrated spectral radiance, line intensity or continuum intensity is measured for lines of sight perpendicular to the axis of the arc (see figure 3.1). The local emission is then often found as a function of the radial coordinate by an Abel transform (see e.g. Lochte-Holtgreven, 1968). To obtain information concerning the radial temperature distribution, a sapphire discharge tube is needed. The scattering that takes place in the conventional polycrystalline alumina is generally too large for these measurements.

The plasma temperature in HPS discharges has been determined by various authors (table 3.5), based on the measurement of one or more of the following seven quantities:

a) Self-Reversed Spectral Lines

The spectral radiance L_m at the intensity maximum of self-reserved non-resonance lines is a sensitive function of the maximum temperature T_m along the line of sight (Bartels, 1950, 1951)

$$L_m = C_B\, B_\lambda\, (T_m) \tag{3.29}$$

where C_B is a constant (so-called Bartels' constant). Knowing the value of C_B, the value of T_m can be derived from the measured spectral radiance L_m via the Planck function B_λ. Bartels has given an analytical approximation for C_B, from which it follows that $C_B \approx 1.5$ for the Na 818.3/819.5 nm lines and $C_B \approx 1.7$ for the Na 568.3/568.8 nm lines. More accurate values for C_B, as a function of plasma temperature distribution and sodium vapour pressure, have been calculated by de Groot (1974b) from a numerical solution of the radiative transfer equation.

Because of its simplicity, this method of temperature measurement seems to be the most appropriate for HPS discharges.

b) Absolute Line Intensities

The emission coefficient ε for a spectral line, integrated over the complete line profile, is proportional to the density n_u in the upper state and the transi-

Table 3.5. *The axis temperature T_A in high-pressure sodium discharges as found by various authors. Xenon is present as ignition gas*

Author(s)	p_{Na} (kPa)	I (A)	R (mm)	T_A (K)	Method
Schmidt (1965)	27	4	3.75	4000	depression of series limit
Schmidt (1965)	27	4	3.75	4640	electrical conductivity
Ozaki (1971b)	13	4	3.7	3210	self-reversed spectral lines
Ozaki (1971b)	27	4	3.7	3090	self-reversed spectral lines
Chamberlain *et al.* (1971)	11	6	4	3950	intensity of strontium lines
Chamberlain and Swift (1972)	16	6	4	3150	intensity of strontium lines
de Groot (1972a)	16	8.5	3.8	4125*	self-reversed spectral lines
de Groot (1974b)	17	5.3	3.8	4000	electrical conductivity
de Groot and van Vliet (1975)	17	5.3	3.8	3850	self-reversed spectral lines
de Groot and van Vliet (1975)	17	5.3	3.8	3700	absolute line intensities
de Groot and van Vliet (1975)	17	5.3	3.8	3900	recombination radiation
Waszink (1975)	36	5.2	3.75	4225	self-reversed spectral lines
Akutsu (1975)	15	4.5	3.8	4270	electrical conductivity
Waszink and Flinsenberg (1978)**	13	8.5	3.75	4120*	Stark broadening K line
Waszink and Flinsenberg (1978)**	26	8.5	3.75	4360*	Stark broadening K line
Wharmby (1984)	8		3.7	4030	self-reversed spectral lines

* Measured at current maximum.
** Relatively high xenon pressure (160 kPa under operating conditions).
p_{Na} is the sodium vapour pressure, I the discharge current and R the inner radius of the discharge tube.

tion probability A (see e.g. Lochte-Holtgreven, 1968)

$$\varepsilon(r) = \frac{1}{4\pi} h \, v_0 \, A \, n_u(r) \tag{3.30}$$

For a line with negligible self-absorption, the plasma temperature can, when the vapour pressure is known, be derived from the local emission coefficient, with the aid of the Boltzmann formula (see Eq. 3.8).

c) Recombination Radiation
The continuous spectrum of the HPS discharge below 408 nm is, for standard conditions, mainly due to recombination radiation. The emission coefficient for this radiation can be calculated, for a known sodium vapour pressure, as a function of temperature (Sec. 3.1). Knowing the sodium vapour pressure, the plasma temperature can now, the other way around, be determined from the absolute intensity of the recombination radiation.

d) Stark-Broadened Spectral Lines

The electron density can be determined from the Stark broadening of spectral lines, the half-width and the shift both being approximately proportional to the electron density. The plasma temperature can now be derived from the electron density, using the Saha equation (Eq. 3.11), if the sodium vapour pressure is known.

e) Depression of Series Limits

The (Stark) broadening of the high-lying levels increases with increasing principal quantum number n. At a certain number n_m there is such an overlap of these strongly-broadened lines and the 'true' continuum of the recombination radiation that this line can no longer be distinguished from the continuum radiation. The value of n_m decreases with increasing electron density. The electron density can now be estimated from n_m with the relation given by Inglis and Teller (1939) (see also Lochte-Holtgreven, 1968) and consequently the plasma temperature can be found via the Saha equation, when the sodium vapour pressure is known.

f) Electrical Conductivity

The electrical conductivity is proportional to the electron density and is therefore strongly dependent upon the plasma temperature (see Sec. 4.1.3). The axis temperature can be determined from the conductance (integral of the electrical conductivity over a cross-section of the discharge), which is equal to the quotient of discharge current and electric field strength, when the sodium vapour pressure and the relative form of the temperature distribution are known.

g) R.F. Thermal Noise

The noise voltage at radio-frequencies was used by Anderson (1975) to determine the temperature in an HPS lamp.

Results

A summary of the axis temperature values as measured by various authors for 'pure' HPS discharges is given in table 3.5. From this table it may be concluded that the axis temperature is about 4000 K for typical HPS discharges ($p_{Na} \approx 15$ kPa, $I \approx 5$ A, $R = 3.8$ mm). This conclusion is confirmed by model calculations (see Sec. 4.2). The measurements of Ozaki (1971b) give extremely low values for the axis temperature (see also Sec. 5.1). The same is the case for the data of Chamberlain and Swift (1972), which are much lower than the data of Chamberlain et al. (1971). The axis temperature decreases with an increase in sodium vapour pressure and increases with an increase in input power as shown in figure 3.24. In practical lamps, where

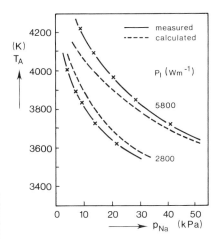

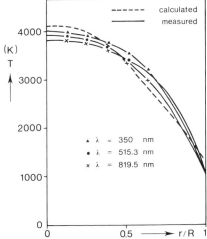

Figure 3.24 Measured and calculated axis temperatures T_A as functions of the sodium vapour pressure p_{Na} for two values of the input power per unit of arc length (P_l). The radius of the discharge tube was 3.8 mm. (de Groot and van Vliet, 1975)

Figure 3.25 Measured and calculated temperature distributions.
The plasma temperature was measured from the recombination radiation at 350 nm, from the self-reversed Na 819.5 nm line, and from the absolute intensity of the Na 515.3 nm line. (de Groot, 1947b). The calculated distribution was obtained with the power balance calculations described in Sec. 4.1. $P_l = 3500\ W\ m^{-1}$, $p_{Na} = 17\ kPa, I = 5.3\ A, R = 3.8\ mm$.

sodium vapour pressure and input power are not independent quantities (as the coldest spot temperature is controlled by the input power), the axis temperature is nearly independent of the input power. This is because of the compensating effects of the increases in sodium vapour pressure and input power.

The temperature distribution lies between a parabolic and a cubic profile (see figure 3.25). The temperature profile becomes increasingly broadened for a higher input power and more constricted for a higher sodium vapour pressure (de Groot, 1974b). This dependence of the plasma temperature distribution on sodium vapour pressure and on input power will be discussed in more detail in Sec. 4.2.1.

Comparison Between a.c. and d.c. Discharges
Unless stated otherwise, the data given for plasma temperature and spectrum are based on time-averaged measurements of the 50 Hz a.c. discharge. The

plasma temperature and the spectrum of the a.c. discharge vary greatly during one cycle of the a.c. supply (de Groot, 1972a, 1972b; Patterson, 1972; de Groot, 1974b; van Vliet and Nederhand, 1977; Rautenberg and Johnson, 1977). Figure 3.26 gives the axis temperature as a function of time within one half-period of the 50 Hz cycle; the moment of current reversal has been taken as time $t = 0$. The axis temperature reaches a minimum about 0.3 ms after the instant of current reversal, while the axis temperature at current maximum is about 1000 K higher than the minimum temperature. This temperature modulation explains the large variations in the spectrum during one cycle of the a.c. supply. Despite this temperature modulation, it may be concluded from measurements on a.c. and d.c. discharges that the plasma temperature and the spectrum as derived from time-averaged quantities for the a.c. discharge (when operated on a sinusoidal current) is a good approximation for the plasma temperature and the spectrum of the d.c. discharge, when comparing discharges with the same input power (de Groot, 1974b; see also Sec. 7.1). However, this conclusion is not valid for pulse-current-operated a.c. discharges, which will be discussed in Sec. 7.2.

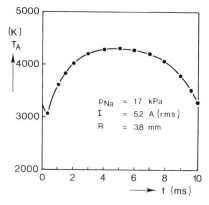

Figure 3.26 Measured variation of axis temperature during a half period of the 50 Hz cycle. (de Groot, 1974b)

3.2.3 Wall Temperature

The temperature of the discharge-tube wall has an important influence on the lifetime and on the luminous efficacy of the HPS lamp (Secs 10.1.2 and 10.1.3). Furthermore, the wall temperature is a boundary condition needed for the measurement or calculation of the plasma temperature distribution. It is therefore important to have a convenient method for determining the wall temperature. Three methods have been used

92

a) Thermocouple Measurements

The wall temperature can be measured with the aid of thermocouples (Akutsu and Saito, 1979). This method suffers from the disadvantage that a special lamp construction has to be employed that allows for the feedthrough of thermocouple wires. Furthermore, thermocouples can disturb the local heat balance.

b) Infrared-Pyrometry

The wall temperature can also be determined, with the aid of infrared-pyrometry, from the spectral radiance of the discharge-tube material in the wavelength region where this material is opaque (so that no discharge radiation is transmitted) and the spectral emissivity is known. It is preferable to use an 8 μm pyrometer as the spectral reflectance of alumina is about zero at this wavelength (figure 8.15), so that the spectral emissivity is approximately unity. A disadvantage of this method is that a special infrared transmitting window has to be made in the glass outer bulb of the lamp under test.

c) D-Line Pyrometry

A method that may be described as D-line pyrometry avoids the disadvantages of the two methods described in the foregoing. With this method the wall temperature is determined from the spectral radiance of the HPS lamp in the wavelength range where the plasma, and not the discharge-tube material, acts as a black body with a temperature equal to that of the wall (de Groot, 1972a). This condition is fulfilled near the centre of the sodium D-lines just

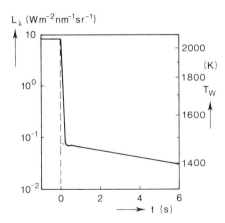

Figure 3.27 Logarithmic registration of the spectral radiance L_λ at 589 nm before and after switch-off of an HPS lamp (switch-off at time $t = 0$). The wall temperature T_W as calculated from L_λ via Planck's function, after correction for reflection due to Al_2O_3 discharge tube and glass outer bulb, is also given.

93

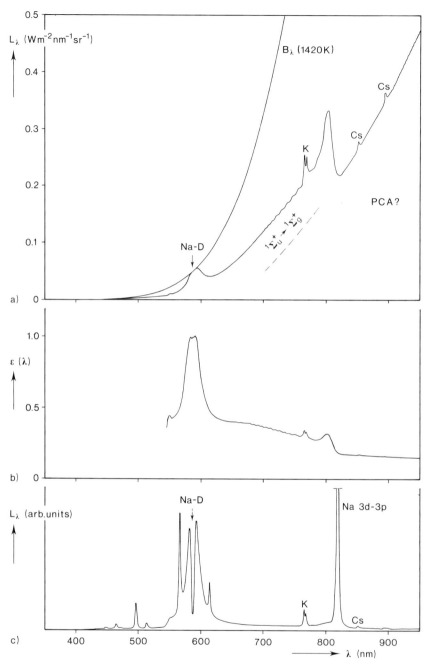

Figure 3.28 (a) 'Afterglow'-spectrum, 0.5 s after switching off an HPS lamp, in comparison with the spectrum $B_\lambda(T)$ of a black body radiator at a temperature $T = 1425$ K. The origin of the HPS lamp radiation is indicated in the figure; part of the infrared continuum above 820 nm is probably due to the polycrystalline alumina (PCA). (b) Spectral emissivity $\varepsilon(\lambda)$ of the afterglow plasma, as determined from the ratio of afterglow-spectrum to black-body spectrum. (c) Emission spectrum of the HPS lamp just before switch-off is given as reference.

94

after switch-off of the lamp. The electron temperature then becomes equal to the gas and wall temperature within milliseconds, the wall temperature decaying over a period of seconds. From a logarithmic plot of the spectral radiance after the moment of switch-off, the value just after switch-off can be found by extrapolation (figure 3.27). After applying a correction for the reflection due to the presence of the discharge tube and the outer bulb, the wall temperature is found from the measured spectral radiance using Planck's law.

The measured spectral radiance closely approximates the spectral radiance of a black body (with a temperature equal to the wall temperature) not only at the centre of the D-lines but over a wavelength range roughly equal to the self-reversal width of the sodium D-lines under operating conditions. This is shown in figure 3.28, where the measured spectrum, about 0.5 s after lamp switch-off, is compared with the spectrum of a black body radiator. The observed 'afterglow'-spectrum resembles the black body spectrum over a larger wavelength range than expected from the calculated atomic spectrum (figure 3.9). This is caused by the fact that the molecular spectrum contributes rather strongly at these low temperatures (see also Wyner and Maya, 1977), in particular for high sodium vapour pressures (high $\Delta\lambda$ values). The wavelength resolution required of the pyrometer is not very high, and this means that an interference filter instead of a monochromator can be used, giving a simple measuring system (figure 3.29).

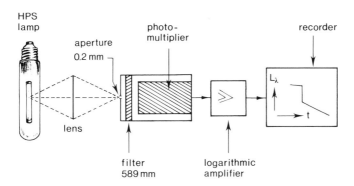

Figure 3.29 Measuring set-up for the measurement of wall temperature using sodium D-line pyrometry.

Because the relevant wavelength is 589 nm, standard production lamps with uncoated glass outer bulbs can be used with this method. D-line pyrometry is a very sensitive method, since a temperature difference of only 10 K already gives a difference in spectral radiance of about 10 per cent, in the temperature

95

range 1300 to 1700 K. A disadvantage of this method is that the lamp has to be switched off for every measuring point. This makes this method more time consuming in comparison with infrared-pyrometry, for the measurement of the wall-temperature distribution, unless an array of detectors is used.

In principle, this method gives the temperature of the plasma, thus of the inner surface of the wall, while the other methods discussed above give the temperature of the outer surface of the wall. The difference between these two temperatures is usually rather small ($\approx$ 10 K), as can be estimated from the heat flow and the thermal conductivity of the wall material.

Figure 3.30 shows measurements of the wall temperature of an HPS lamp as obtained using D-line pyrometry and infrared-pyrometry. It appears that the wall temperature increases approximately linearly with increasing input power. The difference of about $+40$ K found between D-line pyrometry and 8μ-pyrometry is probably due to random errors in the calibration of the pyrometers and not to systematic differences between these two methods, for a similar comparison, carried out by Netten and Kaldenhoven (1976), showed differences of about -20 K.

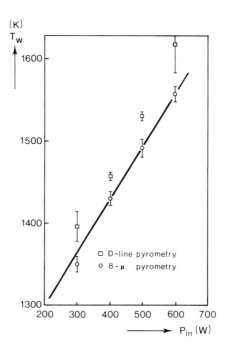

Figure 3.30 Wall temperature T_W, mid-way between the electrodes, as a function of the input power P_{in} for a 400 W HPS lamp. The wall temperature was measured using D-line pyrometry as well as infrared-pyrometry (8μ). (Rickman, 1977)

3.3 Final Remarks

The spectrum of the high-pressure sodium discharge is now fairly-well understood. This holds also for its relation to important lamp characteristics such as chromaticity coordinates, correlated colour temperature, colour rendering and luminous efficacy. The dependence of the spectrum on important discharge parameters like sodium vapour pressure and plasma temperature is also reasonably well known. Conversely information on these discharge parameters can be derived from the measured spectrum.

The spectrum contains many features, such as self-reversed lines, bands, satellites, continua, etc.. Self-absorption and extreme line-broadening play a crucial role in determining the ultimate spectrum. This makes the high-pressure sodium discharge a very interesting object for study from the physical point of view.

The calculated spectra, based on an atomic model for an LTE plasma, give a reasonable description of the experimental spectra for standard high-pressure sodium discharge conditions. At the higher sodium vapour pressures, the molecular contribution should be incorporated to obtain a realistic model.

Note added in proof: an example of such a 'full' calculation of the spectrum, including the molecular contribution, is shown in figure 3.31, in comparison with the calculated atomic spectrum.

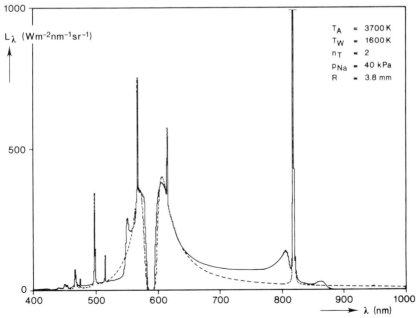

Figure 3.31 Comparison of 'full' calculation of the spectrum, including the $Na_2/NaNa$ contributions (full curve), with calculation of the Na atomic spectrum (dashed curve). (de Groot et al., 1986)

Chapter 4

Power Balance and Related Discharge Properties

For the designer of HPS lamps it is desirable to have a model for the gas discharge. Such a model can help him to obtain a better understanding of the relationships between lamp parameters such as geometry of the discharge tube, the discharge-tube filling and the lamp current on the one hand and lamp properties such as lamp voltage and radiant efficiency on the other. It can also help him to limit the number of experiments needed for optimising lamp design.

It is the power balance equation for the discharge plasma in combination with the associated boundary conditions that relates discharge parameters to discharge properties. The power balance of the total plasma states that the power, acquired by the electrons in the electric field (only a very small part is acquired by the ions), is transferred to the gas atoms in elastic and inelastic collisions between electrons and atoms. Consequently, the gas is heated and radiation generated. The major part of the thermal energy transferred to the gas is passed on to the wall of the discharge tube by conduction. The solution of the power balance equation of the discharge yields the most important discharge properties such as plasma temperature distribution, electric field strength and radiant power for the given discharge parameters.

For the power balance of the complete lamp the electrode power has to be considered. As already shown in Sec. 2.2, the electrode power can be as much as 5 to 20 per cent of the power dissipated in the discharge column. Consequently, to correlate calculated discharge properties with measured lamp properties these so-called electrode losses should be taken into account.

The conclusion was reached in Sec. 3.1 that it is possible to calculate the spectral power distribution for a line of sight perpendicular to the discharge axis from known characteristic data concerning the discharge. These calculations were based on the one-dimensional radiative transfer equation. For the calculation of the local net radiant power in an optically thick non-homogeneous plasma it is necessary to extend the radiative transfer equation to three dimensions. This problem is so complex that – to the authors' knowledge at least – a full calculation of the total radiant power has not yet been given. Simplifications have to be introduced with respect to the radiation transport in order to obtain a practical model as described in Sec. 4.1.

An experimental determination of the various terms in the power balance may provide a check on the validity of such a practical model. Measurements of various discharge properties such as plasma temperature, field strength and radiant power are therefore presented in Sec. 4.2.

In this chapter the high-pressure sodium discharge without buffer gas will be considered. Only a small amount of xenon is added to the discharge tube as a starting gas, because experiments can only be carried out if, besides sodium, a starting gas – mostly xenon at a low pressure – is also present. The influence of buffer gases such as xenon and mercury will be considered in Chapter 5.

4.1 Model Calculations

The various models for HPS discharges reported in literature (Lowke, 1969, 1974; Waszink, 1973, 1975; de Groot and van Vliet, 1975; de Groot et al., 1975a; Waymouth, 1977; Yu-Min and Ling-Tang, 1982; Denbigh et al., 1983; Lee and Cram, 1985) vary greatly in the degree of sophistication in treatment of the basic equations, because of the different purposes the authors had in mind.

From the power balance data for HPS lamps as presented in Sec. 2.5 (figures 2.9–2.11), it is clear that the power balance of such discharges is to a large extent determined by radiative processes. As can be read from the data given in table 2.2, about 85 per cent of the radiant power originates from spectral lines. The continuum radiation amounts to about 15 per cent of the radiant power. The radiation of the sodium D-lines amounts to about 50 per cent of the total radiant power.

Calculations of Lowke (1969), the results of which are shown in figure 4.1, illustrate the importance of radiation and self-absorption of the sodium D-lines for the temperature distribution. Without self-absorption the radiation-dominated sodium discharge would show a constricted temperature profile (figure 4.1a, lower dashed curve). In fact the HPS discharge has a nearly parabolic temperature profile (figure 4.1a, full curve). Such a profile is characteristic for conduction-dominated discharges (figure 4.1a, upper dashed curve). In the radiation dominated HPS discharge this type of temperature profile is caused by the strong self-absorption of the sodium D-lines. The self-absorption causes a marked decrease near the wall of the discharge tube of the net radiant power passing in the outward direction through a cylinder of radius r about the discharge axis (figure 4.1b).

However, models that include only the sodium D-lines underestimate the radiation as the contribution of the non-resonance lines to the net radiant power is about the same as that of these resonance lines (see table 2.2). Although the model of Lowke describes the basic features of the HPS discharge

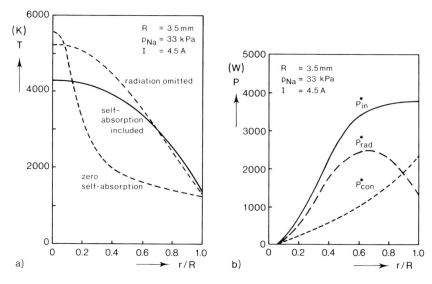

Figure 4.1 The importance of radiation and self-absorption with regard to the temperature profile and the power balance.

(a) Calculated effect of sodium D-line radiation and self-absorption on the temperature profile $T(r/R)$ in HPS discharges. The full line shows the temperature profile for the case where the D-lines are considered as optically thick lines. The dashed lines show the effect of neglecting the self-absorption of the D-lines (zero self-absorption) and of totally neglecting the radiation (radiation omitted) (Lowke, 1969).

(b) The integrals P^*_{rad}, P^*_{in} and P^*_{con} of the net radiant power, input power and conduction power respectively, extended from the discharge axis to the radial coordinate r, as a function of the reduced radial coordinate r/R in an HPS discharge of unit length, based on calculations by Lowke (1969).

very well, his model does not yield data approximating experimental results very well. De Groot and van Vliet (1975) compared results of measurements with the results of two sets of calculations, one of them including the most important non-resonance lines in the model, the other one excluding these spectral lines. They showed that non-resonance radiation has an important influence on calculated plasma temperature and field strength (table 4.1). An adequate treatment of the self-absorption of the non-resonance lines – the Na 568/569, 818/819 and 1138/1140 nm lines are self-reversed at relevant lamp conditions – should be included in the model to obtain an overall agreement with experimental results. For a complete model the continuum and sodium molecular radiation should also be included. However, such a complete model has not, at the time of writing, been described in the literature.

In most models a situation of LTE is assumed. Because a radiating plasma is essentially in a non-equilibrium state, one has to investigate to what extent the assumption of LTE is justified. The investigations of Waszink (1973, 1975) and Akutsu (1975), described in Sec. 3.1.2, lead to the conclusion that

100

Table 4.1 Measured and calculated axis temperature T_A, electric field strength E and power P for a discharge of unit length (electrode losses neglected), to show the influence of the most important non-resonance lines on the mentioned discharge properties. In this case the non-resonance lines were considered as optically thin lines.

Remarks	T_A(K)	E(V m^{-1})	P(W)
measured	3700	1280	4860
only D-lines	4120	780	2960
non-resonance lines included	3770	1450	5510

The measurements and calculations were carried out for a sodium vapour pressure of 33 kPa, a discharge current of 3.8 A and a wall temperature of 1580 K; the diameter of the discharge tube was 7.5 mm. (de Groot and van Vliet, 1975)

LTE may be assumed for the description of the power balance.

With a model in which the plasma temperature distribution is approximated by a two-step temperature profile (Elenbaas, 1951; Waymouth, 1977) – inside the hot core the plasma temperature nearly equals the axis temperature, while outside the hot core it equals the wall temperature (figure 4.2) – simple relations can be derived between some discharge properties and discharge parameters. Although these simplified models do not give such complete and accurate results as the more sophisticated ones, they are undoubtedly of practical use. In Sec. 4.2.1 use will be made of the two-temperature model to explain qualitatively some properties of HPS discharges.

Although HPS lamps are operated from a.c. power sources, most of the time-

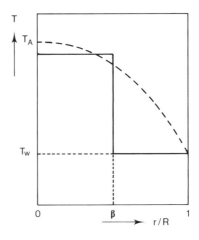

Figure 4.2 Approximation (solid line) of the plasma temperature distribution (dashed line) by a two-step temperature profile. Inside the hot core with radius βR the plasma temperature is constant and nearly equals the axis temperature T_A. Outside the hot core the temperature equals the wall temperature T_W.

averaged properties can be adequately described by a steady-state model. Such a model is dealt with in Sec. 4.1.1. However, the research of the dynamic behaviour of the discharge properties calls for a time-dependent model (van Vliet and Nederhand, 1977) as treated in Sec. 4.1.2. In Chapter 7 use is made of time-dependent models to describe the high-frequency operation of HPS discharges.

As a knowledge of the material properties of sodium plasmas is indispensible when applying the power balance equation to the HPS discharge, Sec. 4.1.3 is devoted to a treatment of such properties. In Sec. 4.1.4 the calculation of the net radiant power emitted by the plasma is explained.

4.1.1 Steady-State Model

The assumption of LTE considerably simplifies the model-based calculations. The conservation equations for the particle densities of the sodium atoms in the various states and of the electrons can be reduced to the Boltzmann and Saha relations respectively – Eqs (3.8) and (3.11). The equation of state relates these densities to the plasma temperature and to the total pressure – Eqs (3.9) and (3.16). Moreover, the assumption of LTE allows the power-balances for electrons and atoms to be reduced to one power-balance equation for both types of particles. This equation states that the power supplied by the electric field to the electron gas σE^2 is partly converted into radiation and partly carried away by thermal conduction. In most cases, convection phenomena need not be taken into consideration (Otani, 1983). The power-balance equation for a volume element in a steady-state HPS discharge is then

$$\sigma E^2 = \operatorname{div}(-\kappa \operatorname{grad} T) + U \tag{4.1}$$

where σ = electrical conductivity
E = electric field strength
κ = thermal conductivity
T = plasma temperature
U = net radiant power (i.e. the difference between radiated and absorbed power) per unit volume

The plasma is assumed to be cylindrically symmetrical and axially homogeneous. Ohm's law relates the electric field strength E and the electrical conductivity σ to the electric discharge current I by

$$I = 2\pi E \int_0^R r\sigma(r)\,\mathrm{d}r \tag{4.2}$$

where R = discharge tube (inner) radius
r = radial coordinate

The boundary conditions are that the first derivative of the temperature at the axis equals zero

$$\frac{dT}{dr} = 0 \quad \text{for } r = 0 \tag{4.3}$$

and that the wall temperature T_W is given

$$T = T_W \quad \text{for } r = R \tag{4.4}$$

To this set of equations may be added a power-balance equation for the calculation of the wall temperature of the discharge tube. The wall temperature will adjust itself such that the power transferred to the wall by the plasma is equal to the power given off to the surrounding medium.

The power-balance equation for the steady-state is a special case of the non-steady case, which will be treated in the next section.

4.1.2 Non-Steady-State Model

In this section a model for the time dependent behaviour of a.c. operated HPS discharges will be described.

For an ideal gas the variation in the quantity of heat per unit volume at radial coordinate r of the plasma can be approximated by

$$\varrho \frac{dq}{dt} = \varrho \left(\frac{dh}{dt} - \frac{1}{\varrho} \frac{dp}{dt} \right) = \varrho c_p \left(\frac{\partial T}{\partial t} + v \frac{\partial T}{\partial r} \right) - \frac{\partial p}{\partial t} - v \frac{\partial p}{\partial r} \tag{4.5}$$

where
q = quantity of heat per unit of mass
h = enthalpy per unit of mass
v = average radial velocity of the particles
ϱ = mass density
p = total pressure
c_p = specific heat capacity at constant pressure

The quantities q, h, v, ϱ, p and c_p are functions of the radial coordinate r. The equations for conservation of mass and momentum have also to be included in such a non-steady-state model, if the effects of radial flow of particles are to be taken into account. Conservation of mass requires such a radial flow to compensate for the variation in time of the densities. The radial flow in HPMV discharges is small, and so too are the pressure gradients as demonstrated by calculations on these discharges (Lowke et al., 1975). The same is assumed for HPS discharges. The variation in the quantity of heat per unit of volume of the plasma can thus be approximated by

$$\varrho \frac{dq}{dt} = \varrho c_p \frac{\partial T}{\partial t} \tag{4.6}$$

103

The time-dependent behaviour can now be described by extending the steady-state power balance equation Eq. (4.1) to include the heating term given in Eq. (4.6), thus

$$\varrho\, c_p \frac{\partial T}{\partial t} = \sigma\, E^2 + \mathrm{div}\, (\kappa\, \mathrm{grad}\, T) - U \tag{4.7}$$

The electric field strength can be calculated by including for the electric current a prescribed function of time in Ohm's law, Eq. (4.2). Alternatively, it can be calculated from an extra equation, the electric circuit equation. For example, in the case of resistance stabilisation this equation can be written as

$$\hat{V}_{m}\cos\omega t = E l + I R_{\mathrm{stab}} + V_{\mathrm{el}} \tag{4.8}$$

where $\hat{V}_{m}$ = amplitude of mains voltage
  ω = circular frequency of supply voltage
  R_{stab} = resistance of lamp stabilising element
  l = distance between electrodes
  V_{el} = electrode fall

At all times the same boundary conditions, Eqs (4.3) and (4.4), have to be satisfied as in the steady-state case. Moreover, the calculation procedure requires the temperature distribution in the discharge to be known at a certain time, considered to be the initial time.

4.1.3 Material Properties

Knowledge of the material properties, viz. electrical conductivity, thermal conductivity and heat capacity, is essential for solving the power balance equation. In this section these material properties will be discussed for a sodium plasma, with xenon as the starting gas.

Electrical Conductivity
Assuming a Maxwellian velocity distribution for the electrons, the electrical conductivity σ can be written as (Shkarofsky *et al.*, 1966)

$$\sigma = n_e\, e\, \mu_e = \frac{4\, n_e\, e^2}{3\, m_e} \int_0^\infty \frac{W^{3/2} \exp\,(-W)\, dW}{v_c(W)} \tag{4.9}$$

with $W = \dfrac{m_e v^2}{2\, k T}$

where n_e = electron number density
   e = elementary charge
   μ_e = electron mobility
   m_e = electron mass
   v_c = collision frequency
   υ = electron velocity
   k = Boltzmann constant
   T = plasma temperature

The collision frequency v_c including the electron-atom and electron-ion interactions is given by

$$v_c = (n_{Na}\, \sigma_{Na} + \sigma_i\, n_i + n_X\, \sigma_X)\, \upsilon \tag{4.10}$$

where σ_{Na}, σ_X = cross-sections for momentum transfer for elastic collisions between electrons and sodium and starting gas atoms respectively
    σ_i = cross-section for momentum transfer for electron sodium-ion collisions
  n_{Na}, n_i, n_X = number densities of sodium atoms, sodium ions, and atoms of the starting gas respectively

The cross-section for momentum transfer for elastic collisions between electrons and sodium atoms is given by Norcross (1971) and Moores and Norcross (1972). The cross-section for momentum transfer for elastic collisions between electrons and xenon atoms is given by Frost and Phelps (1964). The contribution made by electron sodium-ion collisions can be approximated with the formula given by Spitzer (1956)

$$\sigma_i = \frac{4\pi}{\upsilon^4} \left(\frac{e^2}{4\pi\,\varepsilon_o\,m_e}\right)\ln \Lambda \tag{4.11}$$

where ε_o = permittivity in vacuum

and $\Lambda = \dfrac{3(4\pi\varepsilon_o\, kT)^{3/2}}{2\,e^3\,(\pi\,n_e)^{1/2}}$

The expression for the logarithmic factor in Eq. (4.11) is not rigorously correct (Shkarofsky et al., 1966). Minor corrections to the electrical conductivity resulting from electron-electron interactions are therefore not considered.
A reasonable approximation of the electrical conductivity can be obtained by assuming that the collision frequency is independent of the electron velocity. This being the case if the collision cross-section is inversely proportional

to the electron velocity. The electrical conductivity of a sodium plasma can then be approximated by

$$\sigma = \frac{n_e\, e^2}{m_e\, v_c} \qquad (4.12)$$

The collision cross-section for electron-ion collisions is two orders of magnitude larger than for electron-atom collisions. For a degree of ionisation of below a few per cent the electrical conductivity is mainly determined by electron-atom collisions. The electrical conductivity is then proportional to the degree of ionisation

$$\sigma \sim \frac{n_e}{n_{Na}} \qquad (4.13)$$

Above a degree of ionisation of a few per cent the electron-ion collisions become dominant. The electrical conductivity will then be proportional to the ratio of the number density of electrons to that of the sodium ions

$$\sigma \sim \frac{n_e}{n_i} \qquad (4.14)$$

Because of the quasi-neutrality of the plasma, the electrical conductivity becomes temperature independent at high plasma temperatures.

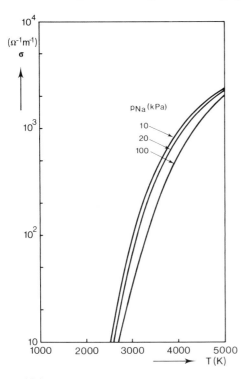

Figure 4.3 Calculated increase in electrical conductivity of sodium vapour with increase in the temperature for three sodium vapour pressures.

Figure 4.3 shows the electrical conductivity of sodium vapour calculated according to relation (4.9), as a function of the temperature for three sodium vapour pressures. The influence of the xenon pressure on the electrical conductivity is given in figure 5.1. If xenon is used as an ignition gas at pressures of about 10 to 30 kPa under discharge conditions, the influence of xenon on the electrical conductivity may be neglected.

The relation between electrical conductivity and plasma temperature can be used to determine the axis temperature – if the relative temperature distribution and the sodium vapour pressure are known – from the current and field strength using Ohm's law as given by Eq. (4.2).

Thermal Conductivity
The thermal conductivity of a vapour or gas like sodium at high temperatures can be calculated using the kinetic theory formula given by Hirschfelder *et al.* (1954)

$$\kappa = \frac{1989.1 \, (T/M)^{1/2}}{\sigma_c^2 \, \Omega^{(2.2)}(T^*)} \tag{4.15}$$

where
$$T = \text{plasma temperature (K)}$$
$$\Omega^{(2.2)} = \text{collision integral for the Lennard Jones potential}$$
$$T^* = kT/\varepsilon = \text{reduced temperature}$$
$$\sigma_c, \, {}^\varepsilon/_\kappa = \text{parameters in potential function (Å and K respectively)}$$
$$M = \text{molar mass}$$

The collision parameters ε/κ and σ_c have been taken from Vargaftik and Voshchinin (1967) for sodium, and from Hogervorst (1971) for xenon. Because of its larger molar mass and collision diameter, the thermal conductivity of xenon is lower than the thermal conductivity of sodium. The influence of the ignition gas xenon on the thermal conductivity of the sodium plasma cannot be neglected, because in HPS discharges with xenon as a starting gas the sodium and xenon pressures during operation are generally of the same order of magnitude.

The influence of the electrons on the thermal conductivity can be examined with the Wiedemann-Franz law relating the thermal conductivity to the electrical conductivity of a slightly ionised gas (Hochstim and Massel, 1969). It appears that the influence of the electrons on the thermal conductivity of the plasma may be neglected for plasma temperatures up to about 4500 K for the gas mixtures found in HPS discharges.

The thermal conductivity of the binary mixture of sodium and a starting gas can be calculated with the help of relations as given by Mason and Saxena (1958). To a good approximation, their theory results in the following simple relation for the thermal conductivity κ of the sodium-starting gas mixture

$$\kappa = \frac{p_{Na}\, \kappa_{Na} + p_X\, \kappa_X}{p_{Na} + p_X} \qquad (4.16)$$

where
p_{Na} = (partial) sodium vapour pressure
p_X = (partial) starting gas pressure
κ_{Na} = thermal conductivity of sodium vapour
κ_X = thermal conductivity of starting gas

Figure 4.4 shows the thermal conductivities of sodium and sodium-xenon mixtures as functions of the temperature calculated according to the relations given in this section. At low Xe-Na pressure ratios the starting gas xenon already has a significant influence on the thermal conductivity of the plasma.

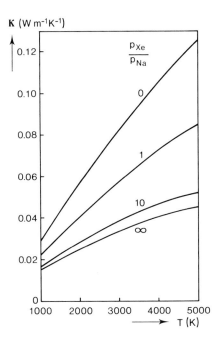

Figure 4.4 Calculated influence of xenon on the thermal conductivity of a sodium plasma as a function of the plasma temperature.

Heat Capacity
The heat capacity of an ideal gas is given by $(5/2)\, k$ where k is the Boltzmann constant. In figure 4.5 the molar heat capacity of sodium used in our model calculations is given as a function of temperature (Stull and Prophet, 1971). The molar heat capacity of sodium plasma approximates the value of $5/2$ kN (N is Avogadro number) as long as excitation and ionisation do not significantly enhance the average energy content of the sodium atom.

108

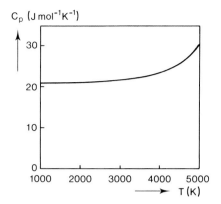

Figure 4.5 The molar heat capacity of a sodium plasma given as a function of the plasma temperature.

4.1.4 Net Radiant Power

The net radiant power of an elementary element of volume in an optically thick plasma is the difference between the local emission of radiation and the local absorption of incident radiation from all other elements of volume of the discharge. Calculation of the net radiant power per unit of volume and per unit of wavelength U_λ at wavelengths for which the plasma is optically thick therefore involves the integration of the local change of the spectral radiance over all the solid angles (figure 4.6)

Figure 4.6 Schematic diagram showing the path of integration, to get the spectral radiance $L_\lambda(r, \bar{s})$ observed in P from direction $\bar{s}$. $\bar{s}$ is a vector with origin in P.

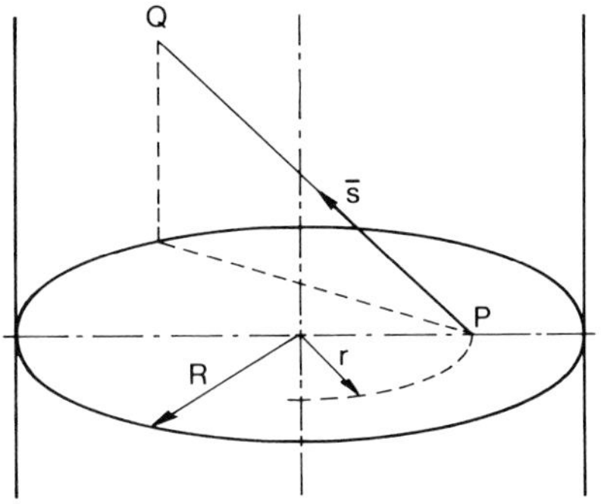

$$U_\lambda(r) = \int_{4\pi} \kappa(\lambda,r)\,[B_\lambda(r) - L_\lambda(r,\bar{s})]\,\mathrm{d}\Omega \qquad (4.17)$$

where $L_\lambda(r,\bar{s})$ = spectral radiance observed in r and coming from direction $\bar{s}$

$\kappa(\lambda,r)$ = absorption coefficient

$B_\lambda(r)$ = black body radiation function

r = radial coordinate

Ω = solid angle

109

The spectral radiance $L_\lambda(r,\bar{s})$ can be calculated by solving the one-dimensional radiative transfer equation (Sec. 3.1) along the path QP illustrated in figure 4.6.

To obtain the net radiant power U per unit of volume of the whole spectrum, an integration over all wavelengths is necessary

$$U(r) = \int_\lambda U_\lambda(r) \, d\lambda \qquad (4.18)$$

It will be clear that the numerical calculation – within an acceptable computer time – of the net radiant power, involving a nest of integrations, is an enormous task. Since the net radiant power is the difference between two large quantities – the emitted and the absorbed radiation – one should be careful to avoid making even relatively small errors in its calculation. This is what makes the calculation of the radiation term in the power balance equation so complicated; it is because of this complexity, and the cost of computation, that this term is generally approximated.

As shown by Holstein (1947), the motion of radiation quanta in an optically thick plasma cannot be regarded as a type of Brownian motion. Because of the highly selective character of the absorption of resonance radiation, it is difficult to define a mean free path length for the motion of resonance quanta. The problem of radiation transfer cannot therefore be treated as a pure diffusion problem. Nevertheless, due to the strong self-absorption of the sodium D-lines, temperature profiles in radiation-dominated, optically thick discharges are very similar to profiles in conduction-dominated discharges, as shown in figure 4.1a.

A description of the radiation transfer in HPS discharges was given by Waszink (1973, 1975). He extended a method, developed by van Trigt and van Laren (1973), for calculating the local net radiant power in the sodium D-lines at a given temperature profile for the case where the ratio $n_1(r)/n_0(r)$ of the densities in the 3p states $n_1(r)$ and the 3s-ground state $n_0(r)$ can be approximated by the following analytical function $f(r)$

$$f(r) = (C_0 + C_2 r^2 + C_4 r^4) \exp(-b\, r^2) \qquad (4.19)$$

C_0, C_2, C_4 and b are constants, which are so adjusted by a least-square method as to give the best fit of $f(r)$ to the function $n_1(r)/n_0(r)$. In this approach, use is made of a radially dependent escape factor $\theta(r)$. This escape factor is defined by the relation

$$U(r) = n_1(r) A_{10} \theta(r) E_1 \qquad (4.20)$$

where A_{10} = transition probability for spontaneous emission for the 3p-3s transition

           $n_1(r)$ = number density of the 3p-states

           E_1 = energy of the 3p-level

For an optically thin plasma $\theta(r) = 1$. The calculation method proposed by van Trigt and van Laren (1973) allows for a relatively fast computation of the escape factor $\theta(r)$ for various functions $f(r)$ as, for example, used above. For a description of the calculation procedure the reader is referred to their publication.

Figure 4.7 shows the by Waszink calculated results for the escape factor and the function $f(r)$, which corresponds reasonably well with the ratio of particle densities $n_1(r)/n_0(r)$. The calculated results presented are part of a model calculation for a pure sodium discharge in which only the D-line radiation is taken into account (Waszink, 1973). This calculation yields an escape factor that is positive for $r/R < 0.65$ and negative for $r/R > 0.65$. This means that the emission exceeds the absorption near the discharge axis and that the absorption exceeds the emission near the wall. The fact that the value for $\theta(r)$ is lower than 0.01 for almost the whole discharge is evidence supporting the statement made above that the emitted and absorbed radiation are both large quantities with respect to the net radiant power.

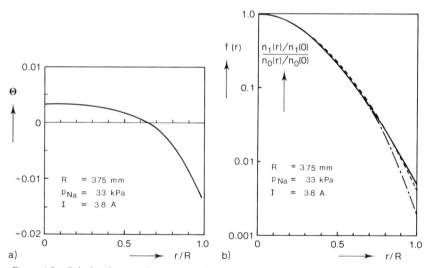

Figure 4.7a Calculated escape factor θ as a function of the reduced radial coordinate r/R of an HPS discharge. (Waszink, 1973)

Figure 4.7b Calculated reduced densities $n_1(r)/n_1(o)$ in the 3p-states with respect to the reduced densities in the 3s-ground-state $n_0(r)/n_0(o)$ as a function of the reduced radial coordinate (full curve). Two analytical functions $f(r)$ are also shown in which the numerical constants are determined so as to give the best fit to the full curve

$----f(r) = [1 + 2.5\left(\frac{r}{R}\right)^4] \exp[-6.76\left(\frac{r}{R}\right)^2]$ (used by Waszink, 1973)

$-\cdot-\cdot-f(r) = \exp[-6.13\left(\frac{r}{R}\right)^2]$ (used by de Groot and van Vliet, 1975)

The calculated results obtained by Waszink are not significantly different from those presented by Lowke (1969), which are based on model calculations using an integral solution of the radiation transfer equation (figures 3.4 and 4.1).

The method of van Trigt and van Laren cannot be applied to the non-resonance lines because it supposes that the density of atoms in the state of the lower energy level as well as the line profile are independent of the radial coordinate. Provided that the inhomogeneity of the plasma is not too severe, this assumption is reasonable for the resonance lines but not for the non-resonance lines (Waszink, 1975).

For the non-resonance lines another approximation can be used, the absorption coefficient of the strongest sodium non-resonance lines is three orders of magnitude lower than that of the D-lines. Thus, in the first place, the influence of self-absorption is much smaller for the non-resonance lines, than for the D-lines. Secondly, the emission and absorption of the non-resonance lines takes place mainly in the hot core of the discharge. Therefore, for the sake of simplicity, de Groot and van Vliet (1975) employed a radially independent escape factor for the non-resonance lines in their discharge model. The value of this escape factor was varied in order to obtain the best overall fit between the results of field strength calculation and measurement for a discharge tube diameter of 7.6 mm and a sodium vapour pressure of 13kPa. This gave an effective escape factor of 0.45 for the non-resonance lines. For the calculation of the net radiant power in the resonance lines they used the same method as Waszink. The ratio of densities $n_1(r)/n_0(r)$, however, is approximated by a simpler function $f(r) = C_0 \exp(-br^2)$ (figure 4.7b) in order to reduce the computation time. This fit of $f(r)$ to the ratio of densities is not very good near the discharge wall, which results in too high values for the calculated radiant power (Sec. 4.2.1). This model is used in this chapter and in Chapters 5 and 7 to obtain a quantitative picture of the influence of certain discharge parameters on various discharge properties.

4.2 Calculated and Measured Results

As stated in the beginning of this chapter, a model can help to provide a better understanding of the relationships between lamp parameters and lamp properties. In Sec. 4.2.1 we will discuss, with the help of steady-state model calculations, the influence of the discharge input power (current), sodium vapour pressure and discharge tube diameter on the plasma temperature, electric field strength and radiant efficiency. The influence of the wall temperature on the discharge properties will be treated separately, in Sec. 4.2.2. Non-steady-state calculated results are presented at the end of this chapter, in Sec. 4.2.3. As a test for the validity of the discharge model presented, certain of the calculated results will be compared with measured discharge properties.

The so-called wall loading P_w is sometimes employed instead of the discharge input power. This wall loading is obtained by dividing the electric input power dissipated in the discharge column (lamp power minus electrode losses) by the inner area of the discharge tube wall between the electrodes

$$P_w = \frac{P_{la} - P_{el}}{\pi D l} \qquad (4.21)$$

where P_{la} = lamp power
        P_{el} = electrode losses
        D = discharge tube (inner) diameter
        l = discharge length between the electrodes

In practice the wall loading should be confined between certain upper and lower limits ($0.10 < P_w < 0.40$ W mm^{-2}). A too-high loading gives rise to an excessively high discharge tube temperature, and this is detrimental to the discharge tube envelope (Chapter 8), while a wall loading that is too low will prevent the build up of a sufficiently high sodium vapour pressure.

The discharges on which most of the experiments described in this section were carried out, were produced in sapphire discharge tubes with inner diameters of 1.8 mm, 4.0 mm, 7.6 mm and 11.2 mm. The separation between the emitter-coated electrodes was 64 mm. The filling was sodium (3 mg), with xenon added as an ignition gas (1.3 kPa at room temperature). The xenon pressure during operation was calculated to be about 10 kPa. The sodium pressure during operation was determined from the value of the wavelength separation between the maxima of the self-reversed D-lines (Sec. 3.2.1). By regulating the temperature of the coldest spot with heat shields, the sodium vapour pressure could be held at the desired value.

4.2.1 Steady-State-Results

Plasma Temperature

In Sec. 3.2.2 measured and calculated temperature distributions were compared, and it was concluded that the agreement between the two was reasonable. Figures 4.8 and 4.9 show the calculated temperature distributions for a wide range of input powers (electrode losses neglected), sodium vapour pressures and discharge tube diameters. From these figures it can be concluded that

a) For a constant sodium vapour pressure, the axis temperature increases by between 200 K and 300 K when the input power is doubled. The temperature distribution then becomes more flattened.

b) For a constant input power, the axis temperature decreases by about 200 K

113

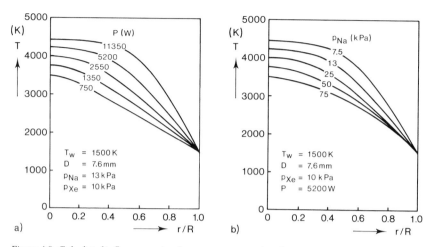

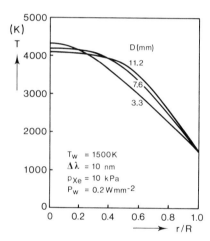

Figure 4.8 Calculated influence on the plasma temperature distribution within a high pressure sodium discharge of (a) the input power per unit of length and (b) the sodium vapour pressure.

Figure 4.9 Calculated influence of the discharge tube diameter on the plasma temperature distribution within a high-pressure sodium discharge.

when the sodium vapour pressure is doubled. The temperature distribution then becomes less flattened.

c) For a constant $\Delta\lambda$ and wall loading, the influence of the diameter of the discharge tube on the axis temperature is small. The temperature distribution flattens with increase in discharge tube diameter.

Application of elementary theory can help to explain the calculated results (Lowke, 1970; Waymouth, 1971). As follows from the power balance equation (Eq. 4.1), the curvature of the temperature distribution at the axis of the discharge is given by

$$\frac{\mathrm{d}^2 T}{\mathrm{d}r^2} = -\frac{1}{2\,\kappa}\,(\sigma E^2 - U) \tag{4.22a}$$

114

and outside the axis by

$$\frac{d^2 T}{dr^2} = -\frac{1}{\kappa}(\sigma E^2 - U) - \frac{1}{r}\frac{dT}{dr} \tag{4.22b}$$

From Eq. (4.22), it is possible to deduce how the curvature of the plasma temperature distribution is influenced by the temperature dependence of the local radiant power and the electric input power.

Because the calculated temperature distributions given in figures 4.8 and 4.9 always exhibit a negative curvature for any value of the radial coordinate, the local input power σE^2 exceeds the local net radiant power U everywhere in the discharge column. As did Elenbaas (1951), we represent the actual sodium atom by an imaginary one having only one excited level with an energy $\bar{E}$ above the ground level, from which radiation occurs by transition to the ground level. Assuming the escape factor for this radiation to be independent of the radial coordinate (as is the case for the sodium D-lines near the centre of the discharge – figure 4.7a) the temperature dependence of the net radiant power can be represented by

$$U = C_1 \, p_{Na} \, T^{-1} \exp(-\bar{E}/kT) \tag{4.23}$$

where C_1 is a constant. Combining the Saha equation, Eq. (3.11), with Eq. (4.13) for the electrical conductivity, the local input power can be given by

$$\sigma E^2 = C_2 \, E^2 \, p_{Na}^{-1/2} \, T^{5/4} \exp(-E_i/2kT) \tag{4.24}$$

where C_2 is a constant and E_i is the ionisation energy of the sodium atom. The temperature dependence of the net radiant power and input power is mainly determined by the exponential factors in Eqs (4.23) and (4.24). Consequently, to get a negative curvature of the temperature profile – the local input and the net radiant power have roughly the same magnitude – the following requirement has to be fulfilled

$$\bar{E} > E_i/2 = 2.57 \text{ eV} \tag{4.25}$$

Conversely, if $\bar{E} < E_i/2$, then one may expect a constricted temperature profile with a partly positive curvature. This in agreement with the constricted temperature profile calculated by considering the sodium D-lines – with their radiating levels at 2.1 eV – as being optically thin (figure 4.1a).

Once the temperature dependence of the net radiant power is known, the temperature profile given for several input powers and sodium vapour pressures can be better understood. If, at constant sodium vapour pressure, the input power is increased, the plasma temperature increases with only a slight change in the field strength (see next heading). Because $\bar{E} > E_i/2$, the net radiant power increases more than the input power, so resulting in a smaller difference between the input power and net radiant power and thus in a more flattened temperature profile. Doubling the sodium vapour pressure results not only in a drop in the plasma temperature, but also in an increase in the field strength (see next section). The combined effect of the higher sodium vapour pressure, the lower plasma temperature and the larger field strength results, for high sodium vapour pressures alone, in a larger difference between the electric input power and the net radiant power, as follows from Eqs (4.23) and (4.24). This is in accordance with the calculated temperature profile for higher sodium vapour pressures, which is steeper.

Electric Field Strength

It is possible to measure the lamp voltage as a function of the input power (current) without varying the sodium vapour pressure. This is done by feeding the discharge with a 'square-wave' current waveform superimposed on a keep-alive current as illustrated in figure 4.10. With such a waveform the current I_{la} can be varied with a constant time-averaged power input by adjusting both current levels simultaneously. Keeping the time-averaged power input constant means a constant sodium vapour pressure. The use of the pulsed current thus allows the measurement of the lamp voltage as a function of the instantaneous current (or input power) at a constant sodium vapour pressure. By changing the polarity every 10 ms, cataphoresis effects are eliminated. Within a few milliseconds after every change of current level a new equilibrium situation will be reached, and the instantaneous lamp voltage is then measured.

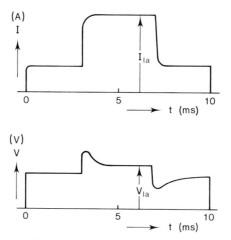

Figure 4.10 Characteristic lamp voltage waveforms obtained by feeding the discharge with a current of 'rectangular' waveform superimposed on a keep-alive current.
V_{la} *is the lamp voltage for lamp current* I_{la} *at time t.*

116

In figure 4.11 the instantaneous lamp voltage measured in this way is given as a function of the instantaneous current value for four sodium vapour pressures for a discharge tube of 7.6 mm internal diameter. From these characteristics it can be seen that at low currents the lamp voltage decreases with increasing current, but that at high current values the lamp voltage increases slowly with increasing current. The same behaviour is found for HPMV discharges as was explained by Elenbaas (1951) from his simplified discharge model. From these characteristics it can be concluded that the lamp voltage is nearly independent of the current except at very low current values.

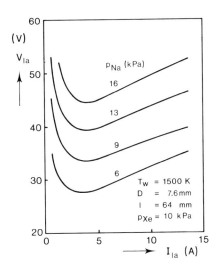

Figure 4.11 V_{la}–I_{la} characteristic of an HPS discharge in which the discharge is fed with a rectangular current waveform as shown in figure 4.10.

Normally, HPS discharges are operated on a sinusoidal discharge current. Because of the negative voltage-current characteristic at low currents, the instantaneous lamp voltage will exhibit the so-called reignition and extinction voltage peaks. The fact that these voltage peaks are smaller than one would expect from the d.c. voltage-current characteristic, is due to the thermal inertia of the plasma (see also Sec. 4.2.3 and Sec. 5.4). The question remains, how can the electric field strength (calculated in a relatively short time, from the steady-state model) be compared with the measured value of the effective lamp voltage. The calculated field strength agrees very well, however, with a 'measured' field strength value E_m derived from the lamp voltage by the following relation

$$E_m = \frac{\alpha_{la}}{0.9} \frac{V_{la} - V_{el}}{l} \qquad (4.26)$$

117

where V_{el} is the electrode fall, and α_{la} is the lamp power factor defined by

$$\alpha_{la} = \frac{P_{la}}{I_{la}V_{la}} \qquad (4.27)$$

Eq. (4.26) gives not only a correction for the electrode fall, but also contains a correction of $\alpha_{la}/0.9$ for the lamp voltage waveform near current zero. For a square-wave lamp voltage waveform and sinusoidal lamp current, α_{la} would be 0.9. Figure 4.12 gives the field strength, as derived from such measurements, as a function of the diameter D of the discharge tube for several values of the wavelength separation $\Delta\lambda$ between the maxima of the self-reversed D-lines. The wall loading was kept at 0.20 Wmm^{-2}. To compare the measured field strength values with the calculated values, use is made of the relation between the sodium vapour pressure and the $\Delta\lambda$ value given in Sec. 3.2.1. As can be seen, at small tube diameters a discrepancy between the calculated and measured field strengths becomes apparent.

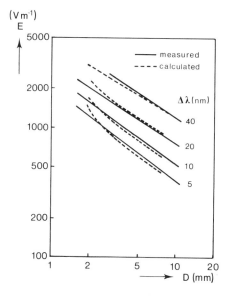

Figure 4.12 Influence of the internal diameter of the discharge tube on the electric field strength for various $\Delta\lambda$-values.

Elenbaas (1951), employing elementary theory, calculated the field strength in an HPMV discharge for various discharge-tube parameters. Such elementary theory is also applicable to the HPS discharge. The most important assumptions made by Elenbaas, can be summarised as follows
- The temperature distribution of the discharge can be described by a two-temperature profile as shown in figure 4.2. Thus the discharge current flows only through the high-temperature part of the discharge having radial co-ordinates $r < \beta R$, were R is the inner radius of the discharge tube.

118

- The net radiant power is proportional to the density of atoms excited to an imaginary level $\bar{E}$, as given in Eq. (4.23).
- The influence of the starting gas is neglected.

With these assumptions he derived the following relation for the electric field strength E

$$E \sim \frac{P^{1/2}}{(P-P^*_{con})^{E_i/4E}} \frac{(p\,D^2)^{1/4\,+\,E_i/4\bar{E}}}{D^{3/2}} \beta^{-1\,+\,E_i/2\bar{E}} \tag{4.28}$$

where
P	=	power dissipated in the discharge column
P^*_{con}	=	conduction power at the edge of the high-temperature core
E_i	=	ionisation energy
$\bar{E}$	=	energy of imaginary level
p	=	vapour pressure of radiating element
D	=	(inner) diameter of discharge tube
β	=	reduced hot-core radius (see figure 4.2)

If one assumes that the diameter of the hot-core is independent of the discharge parameters, and neglecting the electrode fall, then at constant sodium vapour pressure and discharge-tube diameter the lamp voltage is given by

$$V_{la} \sim \frac{P^{1/2}}{(P-P^*_{con})^{E_i/4\bar{E}}} \tag{4.29}$$

At high input powers P^*_{con} can be neglected with respect to P so that

$$V_{la} \sim P^{1/2\,-\,E_i/4\bar{E}} = I^{\frac{1/2\,-\,E_i/4\bar{E}}{1/2\,+\,E_i/4\bar{E}}} \tag{4.30}$$

At high current values the measured electrical characteristic, as shown in figure 4.11, has a positive slope, so the imaginary level $\bar{E}$ will lie above the level $E_i/2$. Hence

$$E_i/2 < \bar{E} < E_i \tag{4.31}$$

This is in accordance with the conclusion reached in the foregoing section, namely that for a non-constricted temperature profile the imaginary energy level $\bar{E}$ must fulfil the requirement (4.31).

The value of the imaginary energy level $\bar{E}$ can be derived from the measured field strength values over a large range of diameters and sodium vapour pressures (figure 4.12). The Elenbaas model predicts how the field strength depends approximately on the diameter and the pressure (see Eq. 4.28). For a constant wavelength separation between the maxima of the self-reversed sodium D-lines, the sodium vapour pressure has to be changed together with the diameter of the discharge tube according to $p_{Na} \sim D^{-1/2}$. The best fit with the measured field strength values given in figure 4.12 is obtained by choosing a value of 4.1 ± 0.4 eV for the imaginary level $\bar{E}$. In the case of a constant $\Delta\lambda$ value, the field strength $E_{\Delta\lambda}$ depends on the discharge tube diameter according to

$$E_{\Delta\lambda} \sim D^{-\frac{9}{8} + \frac{3E_i}{8\bar{E}}} = D^{0.66} \tag{4.32a}$$

The dependence of the field strength on the diameter of the discharge tube and sodium vapour pressure can generally be approximately described by

$$E \sim D^{-0.38} \, p_{\mathrm{Na}}^{0.56} \tag{4.32b}$$

In conclusion, it can be said that the two-step temperature model of Elenbaas can help to explain qualitatively the electrical behaviour of the HPS lamp.

Radiant Efficiency

In figure 4.13 the measured and the calculated radiant powers are given for a discharge of unit length as functions of the input power in the column of the discharge. As can be seen, there exists a linear relationship between the radiant power and the input power. Similar relationships for HPMV and HPS lamps were found by Elenbaas (1951) and Jack and Koedam (1974) respectively. For practical lamps this relationship can be described as follows

$$\begin{aligned} P_{\mathrm{rad}} &= t_{\mathrm{m}} \, (P_{\mathrm{in}} - P_{\mathrm{el}} - P_{\mathrm{con}}^*) \\ &= t_{\mathrm{m}} \, (\pi D l P_{\mathrm{w}} - P_{\mathrm{con}}^*) \end{aligned} \tag{4.33}$$

where
P_{rad} = radiant power
P_{in} = input power
P_{el} = electrode losses
P_{w} = wall loading

The physical interpretation of the constants t_{m} and P_{con}^* can best be understood with the help of figure 4.1b. The electric input power is dissipated in the hot core of the discharge. In the cooler outer mantle, where no power is dissipated, a fraction of the radiant power produced in the hot core is absorbed and converted into conduction power. The constant t_{m} represents the transmission of the cooler outer mantle, the walls of the discharge tube and the outer bulb. P_{con}^* represents the conduction power at the boundary of the hot core.

The radiant power as calculated is higher than that measured with a thermopile. This is due partly to the fact that the absorption of radiation in the cooler outer layer is underestimated in the present model used for the calculations. Another point is that the wall temperature depends on the input power and is not homogeneous along the discharge tube. At high input powers the

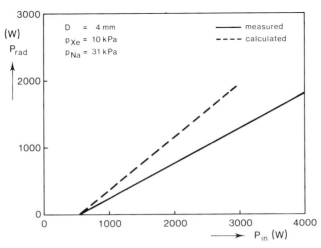

Figure 4.13 Linear relation existing between radiant power and input power per unit of length of the HPS discharge at a constant Δλ-value of 10 nm.

wall temperature of the experimental discharge lamps will be higher and at low input powers smaller than the constant wall temperature of 1500 K used in the calculations. However, the influence of the wall temperature, which will be treated separately in Sec. 4.2.2 cannot explain the large difference between the calculated and measured radiation efficiencies. The absorption of radiation by the discharge tube and by the outer bulb, which has been neglected, is estimated to be about 10 per cent of the radiated power. But the measured results obtained using a thermopile may also possibly lead to too low radiant powers of up to 10 per cent (Sec. 2.3). These are the reasons that for a more quantitive determination of the radiant power, experiments as described in Sec. 2.3 have to be performed.

The measured and calculated radiation efficiencies as functions of the input power show that the radiation efficiency increases with input power and thus with wall loading (see also Eq. 4.33). At a constant discharge-tube diameter and sodium vapour pressure the highest wall loading will therefore yield the highest efficiency.

The influence of sodium vapour pressure (Δλ-value) and wall loading on the radiant efficiency is given in figures 4.14 and 4.15. As shown in figure 4.14, the radiant efficiency increases as Δλ i.e. the sodium vapour pressure increases. This may be explained as follows. As the pressure increases the temperature of the plasma becomes lower, as does the temperature gradient toward the wall. As a result, losses due to heat conduction decrease and the total radiant efficiency slowly increases. For a 'pure' sodium discharge a maximum radiant efficiency of about 55 per cent is possible at a wall loading of 0.2 W mm^{-2}.

121

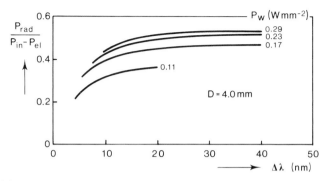

Figure 4.14 Measured influence of the $\Delta\lambda$-value on the radiant efficiency of the HPS discharge for various wall loading values.

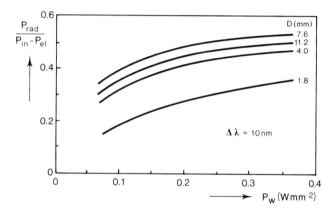

Figure 4.15 Measured influence of the wall loading on the radiant efficiency of the HPS discharge for various discharge tube diameters.

From the experimental data for the radiant efficiency as a function of the wall loading at constant $\Delta\lambda$ as given in figure 4.15, the values for t_m and P^*_{con} can be deduced. From the values for t_m and P^*_{con} as given in table 4.2 it can be concluded that for a constant $\Delta\lambda$-value the transmission t_m of radiation through the cooler outer mantle is nearly independent of the diameter of the discharge tube and that the conduction losses increase only slightly with increase in diameter. As regards t_m, this is because the sodium vapour pressure is so adjusted that the self-absorption of the sodium D-lines does not seriously vary with the tube diameter. For discharges having the same temperature profiles P^*_{con} will obviously be independent of tube diameter. For a temperature distribution as given by Eq. (3.22), P^*_{con} is given by

$$P^*_{con} = -\pi D \kappa \left(\frac{dT}{dr}\right)_{r=R} = 2\pi n_T \kappa (T_A - T_w)^{n_T-1} \qquad (4.34)$$

122

where r = radial coordinate

          D $= 2R =$ inner diameter of discharge tube

          κ = thermal conductivity of the sodium vapour

          T_A = axis temperature

          n_T = power of temperature distribution as given by Eq. (3.22)

          T = plasma temperature

          T_W = wall temperature

Table 4.2 The values for the constants t_m and P^*_{con} for a discharge of unit length derived from the measured relationship between the radiant and input powers according to Eq. (4.33) given for various discharge tube diameters. The $\Delta\lambda$ value is 10 nm.

D(mm)	t_m	P^*_{con} (W)
1.8	0.50	515
4.0	0.52	550
7.6	0.55	650
11.2	0.50	750

Evidently, the discharge tube diameter has dropped out of Eq. (4.34), so that P^*_{con} is, to a first approximation, constant as long as the values T_A, T_W and n_T are the same for various values of the discharge-tube's inner diameter. For not too low values of $\Delta\lambda$ and for constant wall loading the temperature distribution and axis temperature do not strongly depend on the diameter of the discharge tube as is demonstrated in figure 4.9. P^*_{con} is thus to a first approximation independent of the diameter of the discharge tube. The fact that t_m and P^*_{con} are almost unaffected by the tube's diameter means that at constant wall loading and $\Delta\lambda$ the radiant efficiency will decrease for smaller diameters. This can easily be seen from Eq. (4.33). At constant wall loading the input power, and thus also the radiant efficiency decreases for smaller tube diameters. This is why, for low-power lamps with smaller tube diameters, the luminous efficacy of HPS lamps decreases (figure 10.18). For large tube diameters a decrease is measured in the radiant efficiency (at constant $\Delta\lambda$ and wall loading) as well. It is expected that conduction losses to the discharge tube ends then become relevant. For these losses, as for the electrode losses, a correction should be made.

4.2.2 Wall Temperature

With the help of the model presented in this chapter the radiant efficiency of an HPS discharge was calculated as a function of the wall temperature (van Vliet and de Groot, 1982, 1983). From the calculated results given in figure 4.16, it can be concluded that the radiant efficiency shows only a modest increase (1.7 per cent per 100 K increment) at wall temperatures between 1000 K and 2000 K. If the input power or the wall loading is kept constant instead of the lamp current, the calculated increase is approximately 2 per cent per 100 K. The radiant efficiency is equal to unity in the theoretical case that the plasma temperature equals the wall temperature. However, the results for wall temperatures far above 2000 K are of theoretical value only, since such temperatures cannot be realised in experiments.

Calculated temperature profiles as shown in figure 4.17 become flatter with a lower axis temperature as the wall temperature rises. When the difference between the wall temperature and the axis temperature becomes smaller, the heat losses diminish. The lower the axis temperature, the more the excitation of the lower energy levels of the sodium atom is favoured with respect to the higher lying levels. The effect of the relatively enhanced sodium D-line radiation was already shown in figure 3.11, and will give rise to an additional effect on the luminous efficacy by an increase in the useful visual effect of the radiant power. In Sec. 10.1.3 the gain in luminous efficacy so calculated will be checked by experiments.

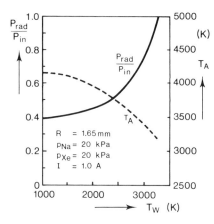

Figure 4.16 Calculated increase in radiant efficiency and decrease in axis temperature of the HPS discharge with increase in the wall temperature of the discharge tube. (van Vliet and de Groot, 1982, 1983)

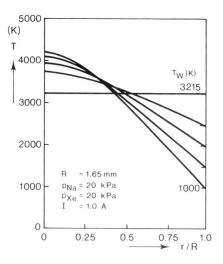

Figure 4.17 Calculated temperature profiles for various wall temperatures. (van Vliet and de Groot, 1982, 1983)

124

4.2.3 Non-Steady-State Results

Because non-steady-state calculations are very time consuming, it is usual to compare steady-state calculated results with measured time-averaged values of discharge properties (Sec. 4.2.1). In this section, however, a few time-resolved solutions of the power balance equation for resistance-stabilised sodium discharges will be compared with the results of measurements to check the non-steady-state model. In the case of resistance-stabilised HPS lamps on a 50 Hz power supply the lamp voltage is dominated by a strong reignition voltage.

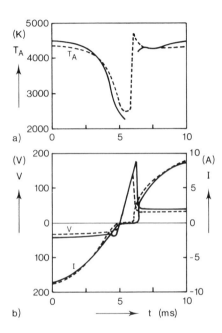

Figure 4.18 Measured and calculated (solid and dotted lines respectively) of (a) axis temperature and (b) lamp voltage as functions of time for a resistive-stabilised HPS discharge. See figure 4.19 for discharge data. (van Vliet and Nederhand, 1977)

In figure 4.18 time-resolved model calculations of current, voltage and axis temperature of the discharge are compared with the measured lamp current, lamp voltage and axis temperature for a resistive-stabilised HPS discharge. The difference between the calculated and the measured voltages can, in the main, be ascribed to the electrode fall (Sec. 9.3), which is not taken into account in the model. From the calculated temperature profiles of figure 4.19 it is seen that reignition takes place through a narrow channel, in which the plasma temperature on the axis reaches high values (see also Sec. 7.2).

Near current zero, the time-dependent power balance for a 'pure' sodium discharge, is dominated by conduction losses. Assuming that the thermal conductivity does not vary with temperature, the time-dependent solution of the power balance equation simplified in this way can be described by

125

a series of Bessel functions of zero order, each of which decays in time. The dominant decay time τ is given by

$$\tau = \frac{\varrho\, c_p\, D^2}{1.4\, \kappa} \tag{4.35}$$

where ϱ = mass density of the sodium vapour
c_p = specific heat capacity at constant pressure
κ = thermal conductivity of the sodium vapour

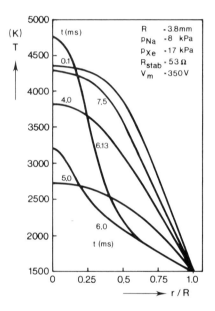

Figure 4.19 Calculated temperature profiles of a resistive-stabilised HPS discharge at various times during one cycle of a 50 Hz power supply with mains voltage $V_m = 350\ V$. The current-zero point is at $t = 5$ ms. (van Vliet and Nederhand, 1977)

For constricted discharges, D stands not for the diameter of the discharge tube but for the effective discharge diameter. For 'pure' sodium discharges at low pressures, the decay time will be very small (0.1 ms). For such small decay times at current zero there is rapid lowering of the plasma temperature to nearly 2000 K. Reignition then takes place through a narrow channel, causing a high axis temperature immediately after reignition. The same effect will be observed if the discharge is operated on current pulses (Sec. 7.2). The modulation of the axis temperature during the period of the supply voltage can be reduced by increasing the relaxation time for plasma cooling using a buffer gas (Sec. 5.4) or by increasing the frequency of the supply voltage (Sec. 7.1).

4.3 Final Remarks

Steady-state and non-steady-state model calculations as presented in this chapter lead to a better understanding of the power balance and related properties of the HPS discharge. This will also be demonstrated in Chapters 5 and 7, where the influence of buffer gases and high-frequency and pulse operation on HPS discharge properties will be treated.

Chapter 5

Influence of a Buffer Gas on Discharge Properties

A gas or vapour added to a sodium discharge in order to influence its properties (including its electrical characteristics, luminous efficacy and colour quality) is called a buffer gas, provided that the spectrum is still dominated by the sodium spectrum. For practical HPS lamps, mercury, or a combination of mercury and xenon at high pressure, is added to the sodium vapour. (A noble gas is always present to make the ignition of the cold lamp possible; this function as an ignition gas – usually at a relatively low pressure of 2 to 4 kPa at room temperature – will be dealt with in Chapter 6.)

A buffer gas may influence the properties of an HPS lamp in various ways. First of all it may change the electric field strength and the plasma temperature (Sec. 5.1) because of changes in the electrical conductivity and the thermal conductivity of the plasma. The spectrum may be changed as a result of an extra broadening of the sodium lines or indirectly by changes in the plasma temperature (Sec. 5.2.). The radiant power and the luminous flux, which are dependent on the changes in plasma temperature and spectrum, will be discussed in Sec. 5.3. Furthermore, the time-dependent behaviour of the a.c. discharge will be influenced when heat capacity and thermal conductivity are significantly changed by the addition of a buffer gas (Sec. 5.4). (The influence that the buffer gas will have on maintenance, lifetime, etc. will be dealt with in Chapter 10.)

As the partial pressures of the buffer gas and the sodium vapour are very important for the properties of the HPS lamp, various methods for determining these partial pressures will be discussed (Sec. 5.5). Sodium forms an amalgam with mercury; the vapour pressure of the sodium as well as that of the mercury depend on the temperature and the composition of the liquid amalgam, as will be discussed in Sec. 5.6.

5.1 Plasma Temperature and Electric Field Strength

The numerical model, described in Chapter 4, with which the power balance equation is solved, can be used to investigate the influence of a buffer gas on the plasma temperature and the electric field strength of an HPS discharge. In these calculations the influence of the buffer gas on the broadening of the sodium lines is neglected. Results will be given for the buffer gases mercury, xenon and helium. Helium is taken as an example to show the effects of a buffer gas with a relatively high thermal conductivity.

5.1.1 Material Properties

For solving the power balance equation, the material properties of the plasma – electrical conductivity, thermal conductivity and heat capacity – have to be known.

Electrical Conductivity
The electrical conductivity of sodium buffer-gas mixtures is calculated using Eqs (4.9) and (4.10), but instead of the starting gas data, the buffer gas data are employed. The cross-sections for momentum transfer for elastic collisions between electrons and the buffer gas atoms are given for xenon by Frost

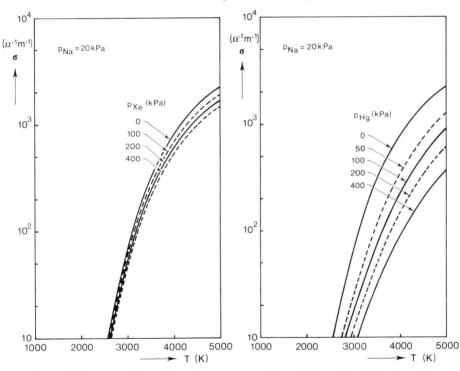

Figure 5.1 Influence of the addition of xenon on the electrical conductivity of sodium vapour.
The calculated electrical conductivity σ is given as a function of temperature T for a constant sodium vapour pressure p_{Na}, at various xenon pressures p_{Xe}.

Figure 5.2 Lowering of the electrical conductivity of sodium vapour by the addition of mercury vapour.

129

and Phelps (1964); for mercury by McCutchen (1958) and Rockwood (1973); and for helium by Gould and Brown (1954). Figures 5.1 and 5.2 show the influence of the addition of xenon and mercury respectively on the electrical conductivity of sodium vapour as a function of the temperature. While the addition of xenon causes only a modest decrease of the electrical conductivity (about 25 per cent for an xenon/sodium pressure ratio $p_{Xe}/p_{Na} = 10$ at a temperature of 4000 K), the addition of mercury results in a considerable reduction of the electrical conductivity (by about a factor of 6 for a mercury/sodium pressure ratio $p_{Hg}/p_{Na} = 10$ at a temperature of 4000 K). The noble gases have only a minor influence on the electrical conductivity (for a given temperature), as the cross-sections for momentum transfer for elastic collisions between electrons and atoms are 1 to 2 orders of magnitude lower for the noble gas atoms than for the sodium atoms. Mercury has a much greater effect on the electrical conductivity, as the cross-sections for momentum transfer for mercury and sodium atoms are roughly the same.

Thermal Conductivity
The thermal conductivity of the sodium buffer-gas mixture can be calculated with the aid of Eqs (4.15) and (4.16). The collision parameters in Eq. (4.15) have been taken from Hogervorst (1971) and from Hirschfelder *et al.* (1954) for xenon and mercury, respectively. The thermal conductivity of helium has been taken from Svehla (1962). The thermal conductivities of sodium-xenon mixtures, as functions of temperature, have already been given in figure 4.4. Figure 5.3 shows the dependence of the thermal conductivity on the relative buffer gas pressure (p_{buf}/p_{Na}) for the case where xenon or mercury is used as the buffer gas. As can be seen, the thermal conductivity is influenced at a relatively low buffer gas pressure. The thermal conductivity is dominated by the buffer gas when the buffer gas pressure exceeds the sodium vapour pressure by at least one order of magnitude. Mercury and xenon cause about the same lowering of the thermal conductivity.

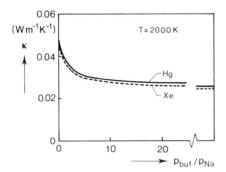

Figure 5.3 Lowering of the thermal conductivity of sodium vapour by the addition of the buffer gases xenon or mercury. The calculated thermal conductivity κ is given as a function of the ratio buffer gas pressure p_{buf} to sodium vapour pressure p_{Na} at a constant temperature.

Mercury and xenon are preferred as the buffer gas in an HPS lamp because of their relatively low thermal conductivities. Table 5.1 gives the thermal conductivities of various noble gases and metal vapours in comparison with that of sodium. The thermal conductivities of the light noble gases, like neon or helium, are much higher than that of sodium and so will cause extra conduction losses when used in an HPS lamp.

Table 5.1 Values of the thermal conductivity κ of various noble gases and metal vapours at a temperature $T = 2000$ K (Svehla, 1962)

noble gas	κ (mW m^{-1} K^{-1})	metal vapour	κ (mW m^{-1} K^{-1})
Xe	24	Na	46
Kr	38	Hg	28
Ar	64	Cd	41
Ne	166	Zn	66
He	540		

Specific Heat Capacity

The specific heat capacity per noble gas atom is given by $\frac{3}{2}k$, where k is the Boltzmann constant (Sec. 4.1.3). This is also approximately valid for mercury at the temperatures considered.

5.1.2 Measured and Calculated Results

Measured and calculated values for electric field strength and axis temperature are given in table 5.2 for HPS discharges with approximately constant values for discharge power per unit of arc length and for sodium vapour pressure, but with various pressures of the buffer gases xenon, mercury and helium. For most cases there is agreement between measured and calculated values, within the uncertainties of the measurements and the calculations (uncertainties in input data and physical model), so it may be concluded that the model gives a satisfactory description of electric field strength and axis temperature in HPS discharges with various buffer gases.

Figure 5.4 shows, in more detail, the influence of the mercury vapour pressure on the axis temperature and the wall temperature of such discharges. These measurements of Wharmby (1984) were performed for constant sodium vapour pressure (controlled by the temperature of an indium bath) and constant input power. The xenon starting gas pressure was very low (0.2 to 1.3 kPa), so that the possible influence of xenon was minimised. The random error

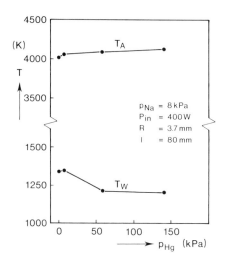

Figure 5.4 *Influence of mercury vapour pressure on axis and wall temperatures of 400 W HPS lamps. The axis temperature T_A and the wall temperature T_W have been measured as functions of the mercury vapour pressure p_{Hg} for constant sodium vapour pressure and lamp input power. The sodium mole fraction of the amalgam has a value between 1 and 0.6 and the xenon starting gas pressure lies between 0.2 and 1.3 kPa, depending on the lamp considered. (Wharmby, 1984)*

Table 5.2 *Comparison of the measured (meas.) and the calculated (calc.) values for electric field strength E and axis temperature T_A for various Na + Hg, Na + Xe and Na + He discharges.*
*The discharge power per unit of arc length ($P_l \approx 4700$ W m^{-1}) is deduced from the measured input power for 50 Hz a.c discharges. The measured values for sodium vapour pressure p_{Na}, buffer gas pressure p_{buf}, current I, wall temperature T_W and arc tube radius (R = 3.8 mm) are used as input data for the **calculations**. The measured values are taken from de Groot (1974b). The 'measured' field strength is derived from the lamp voltage by relation (4.26).*

no	p_{Na}	buffer gas	p_{buf}	I	T_W	E meas.	E calc.	T_A meas.	T_A calc.
	(kPa)	gas	(kPa)	(A)	(K)	(V m^{-1})	(V m^{-1})	(K)	(K)
	$\pm 20\%$		$\pm 20\%$		± 50	$\pm 10\%$		± 150	
1	15	Xe*	10	7.2	1600	725	690	3950	4150
2	15	Hg	80	3.8	1550	1375	1320	4150	4190
3	13	Hg	200	3.3	1550	1580	1700	4300	4270
4	15	Xe	100	6.9	1550	760	750	4050	4060
5	13	Xe	200	6.6	1550	790	730	4100	4080
6	13	Xe	300	6.3	1550	830	760	4000	4070
7	15	He	20	4.28	2000	1220	1140	3700	3850
8	15	He	50	4.05	1950	1300	1380	3600	3800

* Xe is present only as ignition gas

for the measurement of the axis temperature was $\pm$ 60 K. These well-defined measurements show that the axis temperature increases by 100 to 150 K when mercury is added to a sodium discharge, while the wall temperature decreases by some 50 K. The results are in reasonable agreement with those given in table 5.2. The relatively large differences in axis temperature between Na + Hg and Na discharges (about 900 K) as found by Ozaki (1971a, b) are due to the relatively low temperature value found for the Na discharge (Sec. 3.2.2). In the literature, the consensus of opinion is that the measured axis temperature in standard 400 W HPS (Na + Hg) lamps lies between 4100 and 4300 K (Ozaki, 1971a; de Groot, 1974b; Anderson, 1975; Iwai *et al.*, 1977; Otani *et al.*, 1981; Wharmby, 1984).

Measured temperature distributions for HPS discharges with approximately constant values for discharge power and self-reversal width ($\Delta\lambda$) of the sodium D-lines, but with various buffer gases, are given in figure 5.5. Calculated temperature distributions, at constant values of discharge power, sodium va-

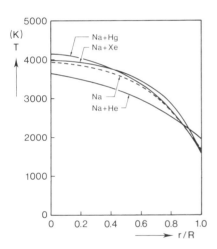

Figure 5.5 Measured temperature distributions for HPS discharges with various buffer gases. The plasma temperature T is given as a function of the normalised radial coordinate (r/R) for Na (1), Na + Hg (2), Na + Xe (6) and Na + He (7) discharges. The number between parentheses refers to the number in table 5.2 where the discharge conditions are given. Discharge power per unit of arc length $P_l \approx 4700$ W m^{-1}, self-reversal width of the Na D-lines $\Delta\lambda \approx 11$ nm.

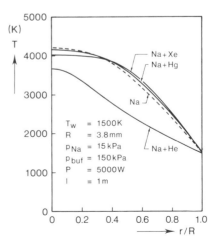

Figure 5.6 Calculated influence of a buffer gas on the temperature distribution in HPS discharges. The calculated plasma temperature T is given as a function of the normalised radial coordinate (r/R) for Na, Na + Hg, Na + Xe and Na + He discharges at constant values of wall temperature T_W, discharge tube radius R, sodium vapour pressure p_{Na} and discharge power P for an arc of length l. The buffer gas pressure $p_{buf} = 10\, p_{Na}$.

133

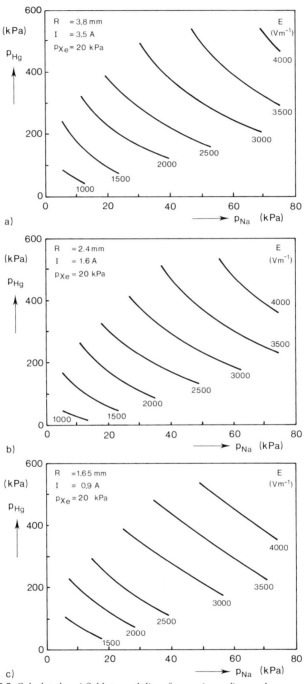

Figure 5.7 Calculated equi field-strength lines for varying sodium and mercury vapour pressures for arc tube radii R of (a) 3.8 mm, (b) 2.4 mm and (c) 1.65 mm. The electric field strength has been calculated from a solution of the power balance equation for constant values of current I and xenon pressure p_{Xe}.

pour pressure and wall temperature, are shown in figure 5.6 for various buffer gases. In these calculations the buffer gas pressure is ten times the sodium vapour pressure. Figure 5.7 shows how the electric field strength increases with increasing mercury vapour pressure; calculated equi-field strength lines are given in this figure for various sodium and mercury vapour pressures and for three values of the arc-tube radius.

From all these calculations and measurements the influence of the various buffer gases can be summarised as follows (for constant sodium vapour pressure and input power)

a) The addition of a heavy noble gas having a relatively low thermal conductivity (like xenon) to a pure sodium discharge results in a broadening of the plasma temperature profile (according to the calculations). This is due to a lowering of the heat conduction losses. The calculations predict a lowering of the axis temperature by about 200 K when a high xenon pressure is added to a pure sodium discharge (figure 5.6). According to the measurements, the axis temperature remains approximately constant while the broadening of the temperature profile is less pronounced than calculated (figure 5.5). It is possible that the actual change in axis temperature at relatively high xenon pressures is smaller than calculated because the influence of xenon on the broadening of the sodium lines (Sec. 5.2) is neglected in the calculations. Increased radiation for which the discharge is optically thin (e.g. in the wings of the sodium D-lines) causes a steeper temperature profile with an increase of the axis temperature (figure 4.1), and this may compensate for the effect that lower conduction losses have on the temperature distribution. The lower conduction losses also result in a lower wall temperature. Due to the relatively small cross-section for momentum transfer in elastic collisions between electrons and xenon atoms, and because of the relatively small changes in effective plasma temperature, xenon has only a minor influence on the electric field strength.

b) The addition of a light noble gas having a relatively high thermal conductivity (like helium) to a pure sodium discharge results in a steeper temperature profile and a decrease in axis temperature because of the accompanying increase in conduction losses. The temperature profile becomes almost linear (figures 5.5 and 5.6). Due to the high conduction losses, the (measured) wall temperature is 300 to 400 K higher than for a Na + Xe discharge with the same input power. Helium causes a strong increase in the electric field strength. This increase is a consequence of the decreased plasma temperature, which results in a strongly-reduced electrical conductivity.

c) The influence of the addition of mercury to a pure sodium discharge is, as far as the plasma temperature is concerned, more complicated than in the case of a noble gas. According to the calculations (figure 5.6), the

135

temperature distributions are approximately the same for Na and Na + Hg discharges. The lowering of the heat conduction losses causes a broadening of the temperature profile and a lowering of the axis temperature (as does xenon), while the lowering of the electrical conductivity increases the axis temperature. According to the calculations both effects compensate each other more or less, so that approximately identical temperature distributions are obtained. According to the measurements, mercury causes an increase in the axis temperature (some 100 to 200 K). The calculations underestimate the change in axis temperature, probably because the influence of mercury on the broadening of the sodium lines is neglected; this broadening may give increased radiation for which the discharge is optically thin. The lower conduction losses, in the case of a Na + Hg discharge, result in a lower wall temperature, which is similar to the effect caused by the addition of xenon. As already mentioned, the addition of mercury causes a strong increase in the electric field strength.

d) The addition of a combination of various buffer gases to a pure sodium discharge is, of course, the most complicated case. For practical HPS lamps the combination of the buffer gases mercury and xenon is important. According to measurements of Iwai *et al.* (1977) and Otani *et al.* (1981), the axis temperature increases while the temperature profile becomes steeper when a high xenon pressure is added to an Na + Hg discharge. This seems to be in contradiction with the effect of xenon in a pure sodium discharge, discussed in the foregoing. However, the discharge conditions are different in the two cases. In the case of an Na + Hg discharge, the thermal conductivity of the plasma will be reduced very little by the addition of xenon while for a pure Na discharge this effect is very important (see figure 5.3). The main effect of adding xenon to an Na + Hg discharge will be to enhance radiation for which the discharge is optically thin. This is due to the extra broadening of the sodium lines (Sec. 5.2) and this may cause a steeper temperature profile, and so a higher axis temperature. As already mentioned, the addition of xenon to an Na + Hg discharge will have only a minor effect on the electric field strength.

5.2 Spectrum

A buffer gas may have a direct influence on the spectrum of an HPS lamp in the form of an extra broadening of the sodium lines, and an indirect influence in the form of a change in the plasma temperature. Emission lines of the additional element will have relatively low intensities as compared to the sodium lines when it has relatively high-lying energy levels compared to the sodium atom and as long as the buffer gas pressure does not exceed the sodium vapour pressure by several orders of magnitude. The latter condition

136

is not fulfilled for the noble gases just after the start of a cold lamp. During the run-up of an HPS lamp with xenon as the ignition gas, the spectrum changes from an xenon spectrum into a sodium spectrum and the xenon lines can then no longer be observed (figure 1.12). The mercury lines are present in a standard HPS lamp, but they are relatively weak under normal operating conditions (figure 1.9).

5.2.1 NaXe and NaHg Spectra

The effects of the buffer gases xenon and mercury on the spectrum are investigated by comparing the measured spectra of sodium discharges having relatively high buffer gas pressures with the spectrum of a so called 'pure' sodium discharge (about 1 kPa xenon is present at room temperature as the ignition gas). This has been done for HPS discharges with an input power of about 150 W.

Xenon
In figure 5.8 the spectrum of an Na + Xe discharge with a relatively high xenon pressure is compared with the spectrum of a 'pure' sodium discharge. For both spectra the $\Delta\lambda$ values (viz. the wavelength separation between the

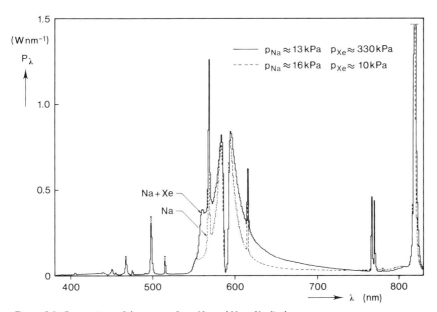

Figure 5.8 Comparison of the spectra from Na and Na + Xe discharges.
The spectral power distributions (1 nm integration interval) were measured in an integrating sphere for 150 W lamps (R = 2.4 mm) with approximately equal wavelength separation between the maxima of the self-reversed sodium D-lines.

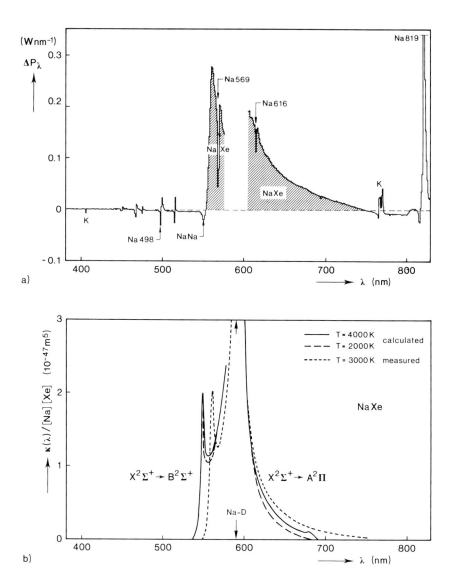

Figure 5.9 Influence of xenon on spectrum.

(a) Difference spectrum for the Na+Xe and Na spectra given in figure 5.8. The spectral power distribution of the Na discharge has been subtracted from that of the Na+Xe discharge: $\Delta P_\lambda = P_\lambda(Na+Xe)-P_\lambda(Na)$. The difference spectrum has been drawn only for wavelength values outside that part of the spectrum where the discharge is optically thick for the radiation of the sodium D-lines. The contribution of the NaXe molecules is indicated by the hatched areas. Various sodium lines are indicated by their wavelengths given in nm.

(b) Calculated absorption spectrum for NaXe molecules. The calculated absorption coefficient $\kappa(\lambda)$, divided by the product of the sodium density [Na] and the xenon density [Xe] is given as a function of wavelength for two values of the temperature (Schlejen and Woerdman, 1985). The transitions are indicated in the figure. The absorption spectrum as derived by Jongerius and Ras (1985) from spectroscopic measurements (T = 3000 K) is also given.

138

maxima of the self-reversed sodium D-lines) are about the same (10 to 11 nm). To show the differences in more detail, the difference spectrum for the two discharges is given in figure 5.9a. The addition of xenon has no strong influence on the intensities of the non-resonant sodium lines in the blue part of the spectrum for which the discharge is optically thin. This result can be explained by the fact that, for a constant sodium vapour pressure, xenon does not have a strong influence on the effective plasma temperature (Sec. 5.1.2). There are indications that xenon, at a high pressure, causes an asymmetric broadening of these sodium lines. The resulting shift of the line centre may cause the 'irregularities' near these lines in the difference spectrum for spectral power distributions measured with a 1 nm integration interval, as given in figure 5.9a. Measurements with higher spectral resolution are needed to clarify these effects. These broadening effects may cause less self-absorption of the non-resonant sodium lines for which the discharge is optically thick, and so could explain the increased power in the Na 819 nm line.

From figure 5.9a it is seen that xenon does have a pronounced influence on the emission in the far wings of the sodium D-lines. It causes a relatively strong and broad band near 560 nm and an enhancement of the red wing of the sodium D-lines for wavelengths up to about 750 nm. This influence of xenon on the broadening of the sodium D-lines was explained by Woerdman and de Groot (1982b). The calculated NaXe absorption spectrum for the Na 3s–3p transition is shown in figure 5.9b; this spectrum was calculated in the modified quasi-static approximation (in the same way as described in Sec. 3.1 for the NaNa spectrum), from the NaXe potential energy curves as published by Laskowski et $al.$ (1981) (figure 5.10). The $A^2\Pi - X^2\Sigma^+$ transition causes an enhancement of the red wing of the sodium D-lines and the $B^2\Sigma^+ - X^2\Sigma^+$ transition an enhancement of the blue wing. A satellite occurs where the potential energy curves are mutually parallel ($dv/dR_{NaXe} = 0$ in Eq. 3.13). This is the case for the $B^2\Sigma^+ - X^2\Sigma^+$ transition at an internuclear distance of about 0.35 nm (figure 5.10), the corresponding wavelength for this situation being about 550 nm (figure 5.9b). This satellite is the band observed for NaXe discharges at 560 nm, the 10 nm difference with the calculated peak wavelength being due to the uncertainties in the (calculated) potential energy curves.

In figure 5.9b the NaXe absorption spectrum as recently derived by Jongerius and Ras (1985) from spectroscopic measurements on Na + Hg + Xe discharges is compared with the calculated spectra. These measurements tend to confirm the good quality of the calculated NaXe spectrum. As the 'measured' spectrum is derived from experimental data, it gives, of course, a more accurate wavelength value for the NaXe satellite. The numerical value of the absorption coefficient for the NaXe satellite is a factor of 5 lower than for the NaNa triplet satellite at 551.5 nm (compare figures 5.9b and 3.8).

This explains why the NaXe satellite can only be observed at relatively high xenon pressures. For standard HPS lamps, where the xenon pressure exceeds the sodium pressure by about a factor of only 2, xenon has a minor influence on the broadening of the sodium lines.

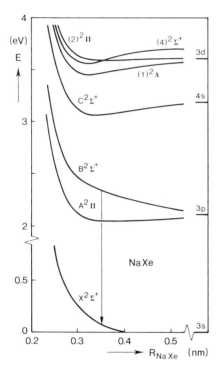

Figure 5.10 Potential energy curves for NaXe.
The energy E of the various levels was calculated as a function of the internuclear separation R_{NaXe} of the Na and Xe atoms (Laskowski et al., 1981). The asymptotic Na levels (limit for large internuclear separation) are identified in the figure. The expected satellite is indicated by an arrow.

The absorption dips present at 569 and 616 nm in the difference spectrum of the Na+Xe and Na discharges (figure 5.9a) are due to the absorption of the Na 568.3/568.8 nm and Na 615.4/616.1 nm radiation (generated in the hot core of the discharge). This absorption is caused by the NaXe (quasi-) molecules present in the cooler outer mantle of the discharge. The NaNa contribution is visible in the difference spectrum (figure 5.9a), because the sodium vapour pressures in the Na and Na+Xe discharges were not exactly the same.

Finally, it should be mentioned that the calculated potential energy curves in figure 5.10 also show an influence of xenon on the high-lying energy levels. This might explain the observed asymmetric broadening of the non-resonant sodium lines, discussed in the foregoing.

Mercury

In figure 5.11 the spectrum of an Na + Hg discharge with a relatively high mercury vapour pressure ($p_{Hg}/p_{Na} \approx 20$) is compared with the spectrum of a 'pure' sodium discharge. In both spectra the $\Delta\lambda_B$-values (viz. the wavelength separation between the centre of the sodium D-lines and the maximum in the blue wing) are about the same (4 to 5 nm) so that the sodium vapour pressures for both discharges are roughly the same (see below). To show the differences between these spectra in more detail, the difference spectrum is given in figure 5.12a. From these figures the following conclusions relating to the addition of mercury can be drawn

– There is an increase in the radiant power of the non-resonant sodium lines, which can be explained by the fact that the axis temperature in the Na + Hg discharge was higher than in the Na discharge. This is due partly to the addition of mercury (as discussed in Sec. 5.1) and partly to a higher input power. The radiant power in the mercury lines is relatively small.

– The NaNa contribution is visible in the difference spectrum. This is because the sodium vapour pressures in the Na and Na + Hg discharges were not exactly the same.

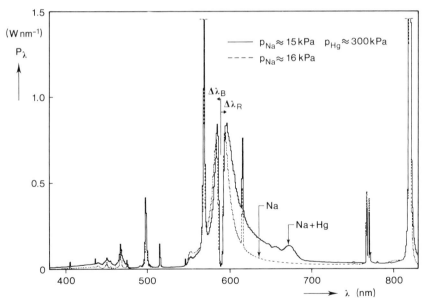

Figure 5.11 Comparison of spectra from Na and Na + Hg discharges.
The spectral power distributions (1 nm integration interval) were measured in an integrating sphere for 150 W lamps (R = 2.4 mm) with approximately constant wavelength separation $\Delta\lambda_B$ between the centre and the maximum in the blue wing of the self-reversed sodium D-lines. The absolute value of the Na + Hg spectrum was reduced by 25 per cent as the input power was 50 W higher than the nominal value.

141

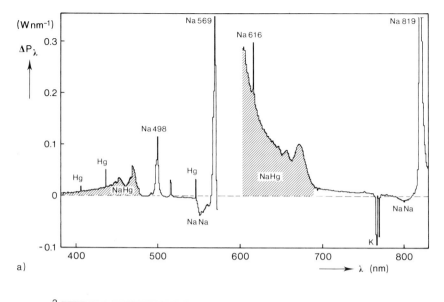

a)

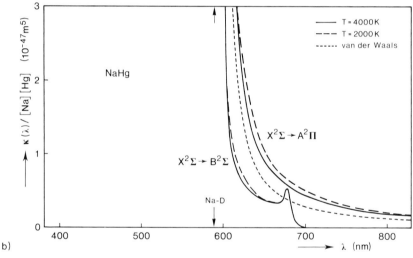

b)

Figure 5.12 Influence of mercury on the spectrum.
(a) Difference spectrum for the Na + Hg and Na spectra given in figure 5.11 The spectral power distribution of the Na discharge has been substracted from that of the Na + Hg discharge: $\Delta P_\lambda = P_\lambda(Na + Hg) - P_\lambda(Na)$. The difference spectrum has been drawn only for wavelength values where the discharge is not optically thick for the radiation of the sodium D-lines. The contribution of the NaHg molecules is indicated by the hatched areas. Various sodium lines are indicated by their wavelengths given in nm.
(b) Calculated absorption spectrum for NaHg molecules. The calculated absorption coefficient $\kappa(\lambda)$, divided by the product of sodium density [Na] and mercury density [Hg] is given as a function of wavelength for two values of the temperature (Schlejen and Woerdman, 1985). The transitions are indicated in the figure. The reduced absorption coefficient based on the van der Waals profile according to Eqs (5.1) and (5.2) is given by the dashed curve ($C_6 = 1.5 \ 10^{-43}$ $m^6 s^{-1}$).

142

- In the blue part of the spectrum a structure arises with peaks at 453 nm and 470 nm, ascribed to NaHg (quasi-)molecules (Woerdman *et al.*, 1985). So far these bands have not been identified. The dependence of these peaks on axis temperature, as observed in time-resolved measurements during the 50 Hz a.c. cycle, suggests that they are due to a transition to the ground state. This NaHg structure in the blue wavelength region may significantly influence the spectrum of an HPS lamp at relatively high sodium and mercury vapour pressures.
- Mercury influences the contour of the sodium D-lines mainly by an enhancement of the red wing, where bands arise at 646 nm, 655 nm and 671 nm. These bands were already observed by Schmidt (1965). The influence of mercury on the red wing of the sodium D-lines will be discussed in more detail in the following.

The fact that the red wing of the sodium D-lines shows an extra broadening when mercury is added to a sodium discharge, while the blue wing remains virtually undisturbed can be explained from the NaHg potential energy curves as published by Hüwel *et al.* (1982), see figure 5.13. Both the $A^2\Pi - X^2\Sigma$ transition and the $B^2\Sigma - X^2\Sigma$ transition have a longer wavelength than the undisturbed Na 3p – 3s transition. The NaHg absorption spectra for these transitions, as calculated by Schlejen and Woerdman (1985) in the modified quasi-static approximation (in the same way as described in Sec. 3.1 for the

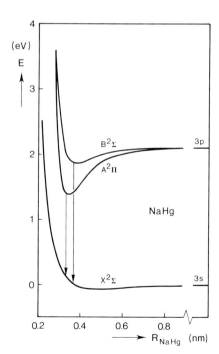

Figure 5.13 Potential energy curves for NaHg. The energy E of the various levels is given as a function of the internuclear separation R_{NaHg} of the Na and Hg atoms (Hüwel et al., 1982). The asymptotic Na levels are identified. The expected satellites are indicated by arrows.

143

NaNa spectrum), are given in figure 5.12b. The $B^2\Sigma - X^2\Sigma$ transition shows a satellite at 679 nm. The experimentally observed bands near 671 nm are identified as this satellite. The measured structure, viz. the three peaks observed at 646 nm, 655 nm and 671 nm, can be explained by a full quantum mechanical calculation, which gives a reasonable description of this satellite structure (Woerdman et al., 1985). The calculated NaHg absorption spectrum for the $A^2\Pi - X^2\Sigma$ transition is rather uncertain. The calculated satellite at 965 nm (outside the wavelength scale of figure 5.12b) is not observed in experimental spectra. This is possibly due to an error in the $A^2\Pi$ potential energy curve given in figure 5.13 (Woerdman et al., 1985).

As a practical approximation, it is often assumed that the extra broadening of the sodium D_1 and D_2 lines can be described as van der Waals broadening with the following profile $W(v,T)$ (de Groot and van Vliet, 1978, 1979; Yu-Min, 1980; Stormberg, 1980; Reiser and Wyner, 1985)

$$W(v,T) = \frac{A_W}{(v_o - v)^{3/2}} \exp\left(\frac{-\pi A_W^2}{v_o - v}\right) , v < v_o \tag{5.1a}$$

$$W(v,T) = 0 \qquad\qquad , v > v_o \tag{5.1b}$$

with $\quad A_W = \tfrac{2}{3}\pi\, C_6^{1/2} n_{Hg}(T)$ $\hfill$ (5.2)

where $\quad W(v,T) =$ van der Waals profile
$\qquad\qquad v_o =$ frequency of the line centre
$\qquad\qquad C_6 =$ constant for van der Waals broadening
$\qquad\qquad n_{Hg} =$ number density of mercury atoms

In the far wing the exponential term in Eq. (5.1a) may be neglected. Representing the profile of the sodium D-lines for Na + Hg discharges by a sum of a Lorentz profile (Na-Na interaction) and a van der Waals profile (Na-Hg interaction) gives a good description of the measured contour in the wavelength range 570 to 630 nm. This is illustrated in figure 5.14 for an Na + Hg discharge with a relatively high mercury vapour pressure, where the extra broadening due to Na-Hg interaction contributes significantly to the broadening of the sodium D-lines. The emission spectrum was calculated from a numerical solution of the one-dimensional radiative transfer equation, looking along a diameter of the discharge (Sec. 3.1.1). For these calculations the value $C_6 = 1.5\,10^{-43}$ m⁶s⁻¹ [see Eq. (5.2)] was used for the effective van der Waals constants (Tiemeyer, 1977). This value is based on a best-fit of contour calculations for Na + Hg discharges with various sodium and mercury vapour pressures (the determination of the vapour pressures is described in Sec. 5.5). The chosen C_6-value (uncertainty about 50 per cent) is in agreement with estimations based on the formula given by Unsöld (1955) (see also Yu-Min, 1980). In recent analyses of the sodium D-line contours in Na + Hg

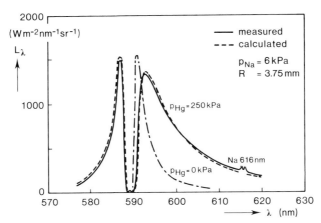

Figure 5.14 Comparison of measured and calculated contours of the sodium D-lines in an Na + Hg discharge with a relatively high mercury vapour pressure. In the calculations of the spectral radiance L_λ the line profile was described by a sum of a Lorentz and a van der Waals profile representing the Na-Na and Na-Hg interactions, respectively. The measurements were performed at the current maximum of a 50 Hz a.c. operated 400 W HPS lamp with a low sodium mole fraction ($x_{Na} \approx .52$) in the amalgam (50 mg). For the calculations, the plasma temperature was taken as being parabolic with an axis temperature $T_A = 5000$ K and a wall temperature $T_W = 1600$ K. The calculated contour for a sodium discharge without mercury ($p_{Hg} = 0$ kPa) is also shown in the figure. (de Groot and van Vliet, 1979)

discharges, values $C_6 \approx 2.8 \, 10^{-43}$ m^{6}s^{-1} (weighted average value for both D-lines) and $C_6 = 1.9 \, 10^{-43}$ m^{6}s^{-1} were found by Reiser and Wyner (1985) and by Jongerius and Ras (1985), respectively.

It should be noted that for the red wing of the sodium D-lines the approximation of the line profile by a van der Waals profile is apparently not valid in the wavelength range extending roughly from 640 to 680 nm where molecular interactions dominate the spectrum. Except for this wavelength range, there is reasonable agreement between the van der Waals approximation for the line profile and the line profile as deduced quasi-statically from the NaHg potential energy curves (figure 5.12b), taking into account the uncertainties in C_6 value and in the $A^2\Pi$ and $B^2\Sigma$ potentials (the $X^2\Sigma - A^2\Pi$ transition in particular has an inaccurately known contribution to the NaHg spectrum).

Miscellaneous

It may be expected that the other noble gases have an influence on the broadening of the sodium D-lines similar to that of xenon. For argon the satellite in the blue wing of the sodium D-lines is somewhat shifted to shorter wavelengths as compared with xenon (Woerdman and de Groot, 1982b).

The influence of cadmium on the spectrum is similar to that of mercury, except for the fact that the bands in the red wing of the sodium D-lines are shifted to longer wavelengths (Mizuno *et al.*, 1971; Zollweg, 1973).

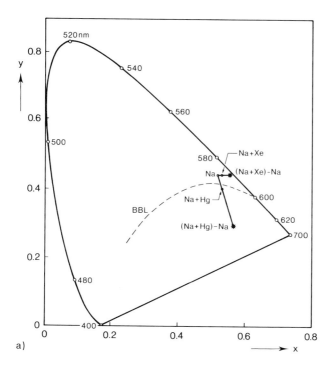

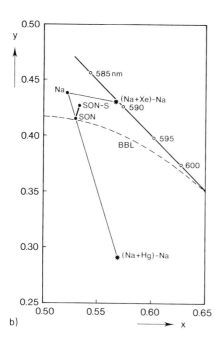

Figure 5.15 Colour points of Na, Na+Xe and Na+Hg discharge lamps in the chromaticity diagram. In those cases where a buffer gas is added to the sodium discharge, the buffer gas pressure exceeds the sodium vapour pressure by a factor of between 20 and 25. The colour points of the difference spectra (Na+Hg)-Na and (Na+Xe)-Na, are indicated in (a) by an asterisk. Figure (b) shows part of the chromaticity diagram on an enlarged scale, where the colour points of the standard HPS lamp (SON) and the HPS lamp with high xenon pressure (SON-S) are given. BBL = black body locus.

5.2.2 Colour and Luminous Efficiency

The changes in the spectrum due to the addition of a buffer gas will have an influence on the colour of the light emitted as well as on the luminous efficiency of the visible radiation (Jacobs and van Vliet, 1980). Figure 5.15a shows the colour points of the spectra given in figures 5.8 and 5.11, viz. for a sodium lamp without buffer gas together with the values for HPS lamps with relatively high xenon and mercury pressures. Also shown in figure 5.15 are the colour points for the difference spectra given in figures 5.9a and 5.12a, obtained by subtracting the spectrum of the sodium discharge, without buffer gas, from the spectra of the discharges with these buffer gases. In fact the colour points of these difference spectra approximately describe the colour of the NaXe and NaHg molecular spectra for a self-reversal width of the sodium D-lines of about 10 nm (there are also changes in non-resonant lines). The colour point of the combination of an Na spectrum and an NaXe or NaHg spectrum now lies on the straight line connecting the colour point of the Na discharge with that of the NaXe or NaHg spectrum.

The addition of mercury causes the colour point to shift toward the red. This can be understood from the influence mercury has on the spectrum, as discussed in Sec. 5.2.1. A limited shift toward the red is desirable in order to obtain a colour point lying nearer to the black body locus than in the case of the 'pure' sodium discharge. The addition of xenon to a sodium discharge causes the colour point to shift toward the yellow region of the spectrum locus. The effect of xenon on the colour point of the sodium discharge is, however, much smaller than that of mercury, as the NaXe spectrum far more closely resembles the NaNa spectrum than does the NaHg spectrum (see Sec. 5.2.1).

Figure 5.16 illustrates how the luminous efficiency of the visible radiation (V_s) is influenced by the addition of a buffer gas. The data given apply for

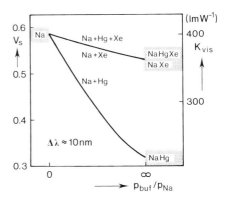

Figure 5.16 Schematic representation of the lowering of the luminous efficiency V_s and the luminous efficacy K_{vis} of the visible radiation by increasing the relative buffer gas pressure. Indicated are the values (hatched areas) for a sodium spectrum (Na) as well as for the NaHg, NaXe and NaHgXe (NaHg + NaXe) spectra. The full lines give the values for combinations of sodium and a buffer gas (Na+Hg, Na+Xe, Na+Hg+Xe). Data apply for a self-reversal width of the sodium D-lines $\Delta\lambda \approx 10$ nm.

a self-reversal width of the sodium D-lines of about 10 nm. The value of V_s for a 'pure' Na discharge is compared in this figure with the V_s values for NaHg and NaXe spectra, as derived from the difference spectra mentioned. The NaHg spectrum has a relatively low V_s value as the radiation is mainly generated in the red, where the spectral luminous efficiency for the human eye (figure 1.9) is relatively low. This explains the rather strong decrease in the luminous efficiency of the visible radiation of an HPS discharge by the addition of mercury (Sec. 10.1.3). The NaXe spectrum has a V_s value, that is not much lower than that of a pure sodium discharge. This explains the fact that the luminous efficiency of the visible radiation of an HPS discharge is only slightly lowered by the addition of xenon (Sec. 10.1.3).

When a combination of two buffer gases (e.g. Hg and Xe) is added to a sodium discharge, an analysis similar to that performed for one buffer gas has to be made to understand its influence on the colour point and the luminous efficiency of the visible radiation. A simple linear combination of the effects of the separate buffer gases does not describe the effect of a combination of buffer gases very well, as the influences of the buffer gases are not always additive. When xenon is added to an Na + Hg discharge there is a (small) colour shift toward the yellow-green spectral region (compare the colour points of SON and SON-S lamps in figure 5.15b) instead of toward the yellow as observed when xenon is added to a 'pure' Na discharge (figure 5.15a). This effect can be understood from the characteristic NaXe and NaHg spectra discussed in Sec. 5.2.1; part of the NaXe spectrum in the red wing of the D-lines will be absorbed by the NaHg molecules, while the NaXe spectrum in the blue wing of the D-lines remains undisturbed.

A similar non-linear effect has been found for the luminous efficiency of the visible radiation. The V_s value for an Na + Hg + Xe discharge is higher than expected for the linear combination of Na + Hg and Na + Xe discharges (for constant values of the self-reversal width $\Delta\lambda$ of the sodium D-lines). For a linear combination of the effects of the NaHg and NaXe spectra one would expect the V_s value of the Na + Hg + Xe discharge to lie between the V_s values for the Na + Hg discharge and the Na + Xe discharge, the actual value depending on the mercury and xenon pressures employed. For practical Na + Hg + Xe discharges it has been found that the V_s value is about equal to the V_s value of Na + Xe discharges, as indicated in figure 5.16. This effect is ascribed to the interaction of NaXe and NaHg spectra for a given self-reversal width of the sodium D-lines.

It is clear from the foregoing that for an accurate description of the spectrum – and consequently of the spectral characteristics of an HPS lamp – a profound investigation into the influence of the buffer gases on line broadening and plasma temperature is necessary. The influence of a buffer gas also depends of course, on the self-reversal width of the sodium D-lines considered. With the aid of numerical calculations it should be possible to quantify these effects reasonably well using the basic data discussed in this section. For these calculations it may be assumed that in the far wings of the sodium D-lines the broadening effects of the various particles are additive as long as binary collisions are dominant.

5.3 Radiant Efficiency and Luminous Efficacy

A buffer gas may influence the radiant efficiency and the luminous efficacy of a high-pressure sodium discharge by changing the power balance of the discharge and the spectral distribution of the radiant power. This is caused by changes in the thermal conduction losses and in the emission and absorption of radiation. Furthermore, the electrode losses may change (at constant input power) when there is a change in lamp current, because the lamp voltage is increased due to the addition of a buffer gas (Sec. 9.3).

The effect of varying the conduction losses is illustrated in table 5.3, where the luminous efficacies of 400 W HPS lamps with various noble gases as the ignition gas are given. The luminous efficacy is lower when a lighter noble gas, with a higher thermal conductivity (see table 5.1), is used.

Table 5.3 The luminous efficacies of 400 W HPS lamps (Na + Hg) with various noble gases as the ignition gas (Beijer et al., 1974b)

Noble gas	Luminous efficacy (lm W^{-1})
Xenon	120
Krypton	116
Argon	110
Neon/argon (99.5%/0.5%)	91

The calculated influence of the buffer gas pressure on conduction losses and radiant power is shown in figure 5.17 for the buffer gases xenon and mercury. The conduction losses decrease when xenon or mercury is added to a pure sodium discharge, and this leads to a higher radiant efficiency. Xenon and mercury have about the same effect on the conduction losses. A significant reduction of the conduction losses may be expected when the buffer gas has roughly the same pressure as the sodium vapour. According to these calculations, conduction losses and radiant power both reach a limiting value when the buffer gas pressure exceeds the sodium vapour pressure by one order of magnitude. As stated earlier, it is assumed with the present model that the buffer gas does not influence the broadening of the sodium lines. As radiation absorption in the discharge tube wall is neglected, the calculated arc efficiencies are too high (Sec. 4.2). Furthermore, electrode losses are not taken into account in these calculations. Because of such limitations, it is not useful to compare the absolute values of calculated radiant power with measurements; only the relative changes are relevant.

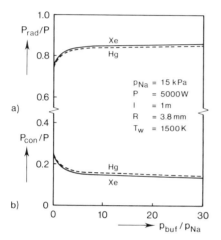

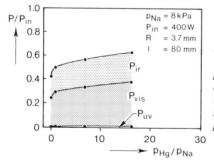

Figure 5.17 Calculated radiant power and conduction losses, relative to the arc input power (P_{rad}/P, P_{con}/P), as a function of the relative buffer gas pressure (p_{buf}/p_{Na}) for the buffer gases xenon and mercury. The calculations are valid for constant values of sodium vapour pressure p_{Na}, discharge power P for a discharge of length l, discharge tube radius R and wall temperature T_W.

Figure 5.18 Measured increase in ultraviolet (P_{uv}), visible (P_{vis}) and infrared (P_{ir}) radiant power, relative to the input power (P_{in}) with increasing p_{Hg}/p_{Na} ratio.
Data apply for 400 W HPS lamps with constant sodium vapour pressure p_{Na} and constant input power P_{in}. (Data after Wharmby, 1984)

Measured data concerning the influence of the buffer gas mercury on the radiation efficiency of an HPS lamp are given in figure 5.18. These experimental data of Wharmby (1984) show a relatively strong increase in the radiant efficiency when the mercury vapour pressure reaches about the same value as the sodium vapour pressure. This is qualitatively in agreement with the calculation results (figure 5.17), so this effect may be ascribed to changes in conduction losses. The measurements further show a steady increase in

150

radiant efficiency with increasing mercury vapour pressure, also at relatively high buffer gas pressures. This effect is not found in the calculations; for high buffer gas pressures no further reduction in conduction losses is to be expected. It is thought that this increase in radiant efficiency at high mercury vapour pressures is due to the extra broadening of the sodium lines at such high pressures. The extra broadening of the sodium D-lines may lead to an increase in the escape factor for the sodium D-line radiation. Less self-absorption will eventually lead to lower conduction losses.

Experimental results for various combinations of the buffer gases mercury and xenon in 150 W HPS lamps are summarised in table 5.4a, while results of power-balance calculations for the same lamps are summarised in table 5.4b. The measured data for the visible and infrared radiation efficiencies

Table 5.4a Measured data for total radiant power (P_{rad}) and radiant power in ultraviolet (P_{uv}), visible (P_{vis}) and infrared (P_{ir}) wavelength regions for experimental 150 W HPS lamps with various buffer gases. The power radiated in the sodium D-lines (P_{Na-D}) is also given as well as the luminous flux (Φ).

Discharge	I_{la} (A)	V_{la} (V)	P_{in} (W)	P_{rad} (W)	P_{uv} (W)	P_{vis} (W)	P_{ir} (W)	P_{Na-D} (W)	Φ (lm)
Na	3.04	60	150	62.5	0.2	32.6	29.7	30	13040
Na + Hg	1.32	132	150	74.1	0.5	41.0	32.0	35	15160
Na + Xe	2.92	56	150	76.5	0.2	45.5	30.8	43	17530
Standard HPS	1.82	97	150	72.2	0.5	39.8	31.9	35	15540
Standard HPS + Xe	1.64	104	150	80.6	0.4	47.4	32.8	43.5	18300

The wavelength separation between the maxima of the self-reversed sodium D-lines ($\Delta\lambda$) has a value between 7 and 11 nm. Arc tube inner radius $R = 2.4$ mm; electrode spacing $l = 58$ mm.

Table 5.4b Calculated results for 150 W HPS discharges.
Electric field strength E, arc power P, conduction power P_{con}, radiant power P_{rad}, Na D-lines radiation P_{Na-D} and axis temperature T_A have been calculated for given current I and partial pressures of sodium (p_{Na}), xenon (p_{xe}) and mercury (p_{Hg}). Power values are given for an arc length $l = 58$ mm. Arc tube radius $R = 2.4$ mm.

Discharge	p_{Na} (kPa)	p_{Xe} (kPa)	p_{Hg} (kPa)	I (A)	E (V m^{-1})	P (W)	P_{con} (W)	P_{rad} (W)	P_{Na-D} (W)	T_A (K)
Na	16	10	0	3.04	828	146	54.6	90	31.5	4290
Na + Hg	8	27	200	1.32	1647	125	42.2	82	22.6	4700
Na + Xe	13	330	0	2.92	886	150	39.3	112	38.1	4310
Standard HPS	12	30	100	1.82	1447	152	42.3	109	32.6	4550
Standard HPS + Xe	12	300	100	1.64	1494	142	40.6	105	33.3	4520

along with the relative luminous efficacy are presented in graphical form in figure 5.19 for the 'pure' sodium discharge (Na) in comparison with the standard HPS lamp (Na + Hg) and the standard HPS lamp with high xenon pressure (Na + Hg + Xe). This is to make more clear the influence of the buffer gas mercury and the combination of mercury and xenon on these parameters. These data show that the addition of the buffer gases xenon and mercury may significantly improve the radiant efficiency as well as the luminous efficacy of a high-pressure sodium discharge. The addition of the buffer gas mercury or the combination of mercury and xenon causes a stronger increase of the visible radiation (dominated by the sodium D-lines) than of the infrared

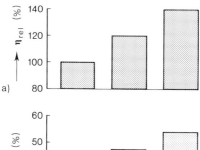

a)

Figure 5.19 Measured data for (a) relative luminous efficacy η_{rel} and (b) visible and infrared radiant efficiency, for experimental 150 W HPS lamps. Data for the 'pure' sodium (Na) discharge are compared with data for the 'standard' HPS (Na + Hg) lamp and the standard HPS lamp with high xenon pressure (Na + Hg + Xe).

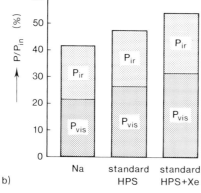

b)

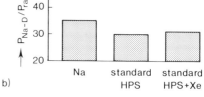

a)

b)

Figure 5.20 Comparison of (a) measured and (b) calculated ratios of sodium D-line radiation P_{Na-D} to total radiation P_{rad} for experimental 150 W HPS lamps. Data for the 'pure' sodium (Na) discharge are compared with data for the standard HPS (Na + Hg) lamp and the standard HPS lamp with high xenon pressure (Na + Hg + Xe).

radiation (mostly non-resonant lines), see figure 5.19. The measurements show that compared to the 'pure' sodium discharge, the ratio of sodium D-line radiation to total radiation remains approximately constant for a high mercury vapour pressure and increases for a combination of high mercury and xenon pressures, while the calculations predict a decrease in this ratio for both cases because of the higher axis temperatures (figure 5.20, table 5.4b). As already mentioned, the increased sodium D-line radiation at high buffer gas pressures is probably due to the extra broadening of the sodium D-lines. For the standard HPS (Na + Hg) lamp this effect is such that it apparently compensates for the expected decrease in the ratio sodium D-line radiation to total radiation at higher axis temperatures. We conclude that the higher efficiency for generation of visible radiation in the case of the standard HPS lamp, as compared with the pure sodium lamp, is due partly to the lowering of the conduction losses and partly to the extra line broadening.

When adding a high xenon pressure to a standard HPS (Na + Hg) lamp, the sodium D-line radiation increases mainly by the effect xenon has on the line broadening, the resulting changes in effective plasma temperature and conduction losses being rather small. We conclude that the improvement in radiant efficiency and luminous efficacy of HPS lamps with high xenon pressure (Na + Hg + Xe), as compared with standard HPS lamps (Na + Hg), is due mainly to changes in the broadening of the sodium lines and not to a reduction of the thermal conduction losses as is generally assumed (Schmidt, 1966; Iwai *et al.*, 1977; Otani *et al.*, 1981, 1982).

5.4 Time-Dependent Behaviour

In Sec. 4.2.3 the conclusion was reached that the modulation of the plasma temperature during one period of the a.c. operated discharge is influenced by the relaxation time for plasma cooling. According to Eq. (4.35) a buffer gas influences this relaxation time. This is due to the larger heat capacity (caused by the increased mass density) and to changes in thermal conductivity of the plasma or radius of the hot channel. The addition of the buffer gases xenon or mercury, or both together, therefore produces an increase in the relaxation time. Due to the increased relaxation time, the a.c. discharge will show smaller temperature variations than does the pure sodium discharge. This is shown in figure 5.21 where the calculated time-dependent behaviour of an Na discharge is compared with the time-dependent behaviour of an Na + Xe discharge during one half-period of a 50 Hz a.c. discharge. As a consequence of the longer relaxation time of the Na + Xe discharge, the reignition and extinction peaks in the field strength of the a.c. Na discharge are reduced. A similar, but less-pronounced effect is produced by the addition

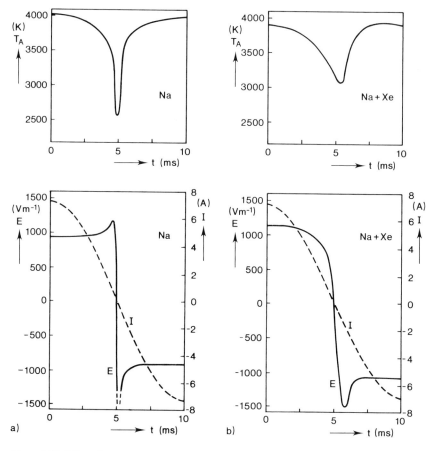

Figure 5.21 Effect of xenon on temperature modulation.
Calculated field strength E and axis temperature T_A are given as functions of time for HPS discharges operated on sinusoidal current (a) without xenon, and (b) with xenon at high pressure ($p_{Xe} = 230 \, kPa$). Arc tube radius $R = 3.8 \, mm$, sodium vapour pressure $p_{Na} = 17.5 \, kPa$, r.m.s. value for current $I = 4.45 \, A$. (de Groot et al., 1975a)

of a high xenon pressure to a standard HPS (Na+Hg) lamp (Iwai et al., 1977; Anderson and Miles, 1982).

Above a certain relaxation time, a self-stabilising discharge (zero ballast impedance) is possible (van Vliet and Nederhand, 1977). This is because the electrical conductance during the heating cycle increases so slowly that the supply voltage is already decreasing before the current can rise to a destructive value. The long relaxation time, necessary for a self-stabilising discharge, can be achieved by the application of a relatively high xenon pressure and by increasing the arc tube radius. Measured and calculated characteristics of such a self-stabilising discharge are given in figure 5.22. The practical application of such a self-stabilising discharge has not yet been possible because of the associated problems during ignition and the changes in lamp characteristics caused by mains voltage fluctuations.

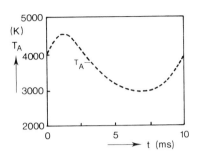

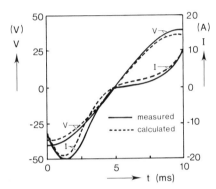

Figure 5.22 Measured and calculated values for axis temperature T_A, lamp and arc voltage V, and current I for a self-stabilising high-pressure sodium discharge with xenon as the buffer gas. The lamp voltage is equal to the mains voltage ($V_m = 28$ V, r.m.s. value), $p_{Na} = 4$ kPa, $p_{Xe} = 230$ kPa, R = 5.6 mm. (van Vliet and Nederhand, 1977)

5.5 Determination of Gas and Vapour Pressures

For a correct understanding and description of the discharge it is essential to know the sodium as well as the buffer gas pressure under actual operating conditions.

5.5.1 Noble Gas Pressure

The noble gas pressure can be determined from the filling pressure at room temperature together with the radial and axial plasma temperature distributions, the total number of noble gas atoms remaining constant. For typical HPS lamp conditions the noble gas pressure during operation is about eight times higher than at room temperature.

155

5.5.2 Sodium Vapour Pressure

The sodium vapour pressure can be determined with the following two methods

a) From the temperature of the liquid sodium at the coldest spot in the discharge volume (indium bath method, as described in Sec. 3.2.1).

When a noble gas is added as a buffer gas, this method may be the same as that used for a pure sodium discharge. However, when mercury is employed as the buffer gas, the sodium and the mercury form a sodium amalgam, so that the sodium vapour pressure only can be determined from the amalgam temperature when the amalgam composition is known (see Sec. 5.6).

b) From the wavelength separation $\Delta\lambda$ between the self-reversal maxima of the sodium D-lines.

When a noble gas is added as the buffer gas, this method is valid only so long as the noble gas pressure is not too high (up to about an order of magnitude higher than the sodium vapour pressure), so that the broadening of the sodium D-lines is not seriously affected by the noble gas present. With the help of the data for the NaXe absorption spectrum, presented in Sec. 5.2, the line contour of the sodium D-lines in an Na + Xe discharge can be calculated from a numerical solution of the radiative transfer equation (Sec. 3.1.1). In this manner the influence of the NaXe emission and absorption on the self-reversal width of the sodium D-lines can be evaluated. From such calculations it is concluded that for a $\Delta\lambda$ value of about 10 nm this $\Delta\lambda$ value increases by about 10 per cent when the xenon pressure is a factor of 20 higher than the sodium vapour pressure (Jongerius and Ras, 1985).

When mercury is added as the buffer gas, method (b) can only be used if a correction is applied for the extra broadening of the sodium D-lines caused by the mercury atoms. Such corrections may be based on calculations of the sodium D-line contour using the line-broadening data discussed in Sec. 5.2. As this extra broadening is mainly in the red wing of the D-lines, and the blue wing remains virtually undisturbed, Eq. (3.26) can still be used to obtain a first-order approximation of the sodium vapour pressure by setting

$$p_{Na} = C_1^{-1/2} \, 2\Delta\lambda_B \, R^{-1/2} \left[\int_{-1}^{1} \frac{1}{T(x)^2} \, dx \right]^{-1/2} \tag{5.3}$$

where $\Delta\lambda_B$ is the wavelength separation between the centre of the D-lines and the self-reversal maximum in the blue wing.

5.5.3 Mercury Vapour Pressure

Several methods have been used for determining the partial vapour pressure of the buffer gas mercury

a) From the amalgam temperature.

When the amalgam composition is known, the mercury vapour pressure can be determined from the amalgam temperature (indium bath method); the relations between mercury vapour pressure, amalgam temperature and amalgam composition are given in Sec. 5.6.

b) From the measured electric field strength.

When the sodium vapour pressure is known (e.g. from the self-reversal width of the sodium D-lines), the mercury vapour pressure can be determined from the measured field strength using the calculated relations between field strength, sodium vapour pressure, mercury vapour pressure and discharge-tube diameter as discussed in Sec. 5.1.2. The use of this method will be discussed in Sec. 10.1.2; it is very useful for practical lamps.

A similar technique is based on known relations between field strength, self-reversal width of the sodium D-lines, amalgam composition and amalgam temperature (Zollweg and Kussmaul, 1983). These relations are derived from a calibration using lamps with known amalgam composition and amalgam temperature. With this technique the mercury vapour pressure is not in fact determined direct, but via amalgam composition and amalgam temperature.

c) From the self-reversed Hg 254 nm line.

The contour of the self-reversed mercury resonance line at 254 nm gives information on the mercury vapour pressure, just as do the self-reversed sodium resonance lines for sodium. The line profile of the long-wavelength wing of the Hg 254 nm line can be described by a van der Waals profile as given in Eq. (5.1) (Perrin-Lagarde and Lennuier, 1971; Laporte and Damany, 1979). In a way similar to that described in Sec. 3.2.1 for the sodium D-lines, the following relation between mercury vapour pressure p_{Hg} and wavelength difference $\Delta\lambda_L$ between line-centre and long-wavelength self-reversal maximum can be derived

$$\Delta\lambda_L = C_{Hg}\, p_{Hg}^{4/3}\, R^{2/3} \left[\int_0^1 \frac{1}{T(x)^2}\ dx \right]^{2/3} \tag{5.4}$$

where C_{Hg} is a constant and x is the normalised radial coordinate.

The relation between $\Delta\lambda_L$ and the mercury vapour pressure p_{Hg} was determined experimentally by de Groot and van Vliet (1978) from measurements on high-pressure mercury discharges (without sodium) in sapphire tubes (figure 5.23). The discharge tubes were mounted in quartz outer bulbs transmitting 254 nm radiation; the mercury vapour pressure in these discharges was known from the mercury filling and the plasma temperature.

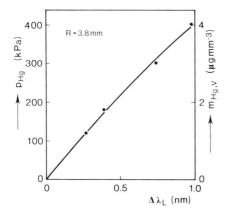

Figure 5.23 The relation between mercury filling per unit of discharge tube volume $m_{Hg,V}$ and wavelength separation $\Delta\lambda_L$ between line centre and reversal maximum in the long-wavelength wing of the Hg 254 nm line. The relation between $m_{Hg,V}$ and mercury vapour pressure p_{Hg} is valid for 400 W HPS lamps with arc tube radius $R = 3.8$ mm and a parabolic temperature profile (axis temperature $T_A = 4500$ K, wall temperature $T_W = 1500$ K).

d) From the extra broadening of the sodium D-lines.
The extra broadening of the sodium D-lines due to the mercury atoms also gives information on the mercury vapour pressure (de Groot and van Vliet, 1979; Yu-Min, 1980; Reiser and Wyner, 1985). As was explained in Sec. 5.2.1, the contour of the D-lines in an Na + Hg discharge can be described in the far wings by a sum of a Lorentz profile (due to Na-Na interaction) and a van der Waals profile (due to Na-Hg interaction). The latter influences only the red wing. The absorption coefficient $\kappa(v,T)$ is given by the following relations

$$\kappa(v,T) \approx C_1 \frac{p_{Na}^2}{T(r)^2} \frac{1}{(v-v_0)^2} \tag{5.5a}$$

for the blue wing of the Na D-lines ($v > v_0$) and

$$\kappa(v,T) \approx C_1 \frac{p_{Na}^2}{T(r)^2} \frac{1}{(v_0-v)^2} + C_2 \frac{p_{Hg}\,p_{Na}}{T(r)^2} \frac{1}{(v_0-v)^{3/2}} \tag{5.5b}$$

for the red wing of the Na D-lines ($v < v_0$), where C_1 and C_2 are constants. As discussed in Sec. 3.1, the contour of the sodium D-lines can be calculated from a numerical solution of the radiative transfer equation; by fitting the calculated contour to the measured contour, using sodium and mercury vapour pressures as variable parameters, these pressures can be determined.

158

A simplified method is based on similar calculations, but considers only the self-reversal maxima. To a first-order approximation, a relation can be found between the asymmetry in the self-reversal widths of the blue and the red wings and the ratio between mercury and sodium vapour pressures. At the self-reversal maxima the optical depth of the discharge is the same for both the blue and the red wings. This leads to the following relation, when considering the sodium D-lines as a single line and converting frequency into wavelength

$$\left(\frac{\Delta\lambda_R}{\Delta\lambda_B}\right)^2 \approx 1 + C \frac{p_{Hg}}{p_{Na}} \Delta\lambda_R^{1/2} \tag{5.6}$$

where $\Delta\lambda_R$ = the wavelength separation between the centre of the D-lines and the self-reversal maximum in the red wing

$\Delta\lambda_B$ = the wavelength separation between the centre of the D-lines and the self-reversal maximum in the blue wing

C = constant

From measurements on 350 W HPS arc tubes with various amalgam compositions and amalgam temperatures (and hence various sodium and mercury vapour pressures) as summarised in figure 5.24, the following practical formula for the determination of the ratio between mercury and sodium vapour pressures was formulated by de Groot and van Vliet (1979)

$$\frac{p_{Hg}}{p_{Na}} = \frac{\left(\frac{\Delta\lambda_R}{\Delta\lambda_B}\right)^2 - 1}{0.057 \, (\Delta\lambda_R^{1/2} - 0.85)} \tag{5.7}$$

where $\Delta\lambda_R$ and $\Delta\lambda_B$ are expressed in nm. This formula differs from the analytical approximation in Eq. (5.6) in so far that there is an additional term present in the denominator, namely -0.85, which is due to the fact that there are the two D-lines (with 0.6 nm separation), whereas for the derivation of Eq. (5.6) the presence of a single line was assumed. The accuracy in the ratio of mercury to sodium vapour pressure as determined with Eq. (5.7) is estimated to be 20 to 30 per cent in the range of sodium and mercury vapour pressures tested. Application of Eq. (5.7) will only give reliable results if the mercury vapour pressure is sufficiently high ($p_{Hg}/p_{Na} > 10$) to give a significant shift of the red wing maximum.

e) From the ratio of mercury and sodium line intensities.

The ratio between mercury and sodium vapour pressures can also be determined from the intensity ratio between Hg and Na lines when the plasma temperature is known. This method was applied by van den Hoek and Visser (1981) using laser-induced fluorescence spectroscopy.

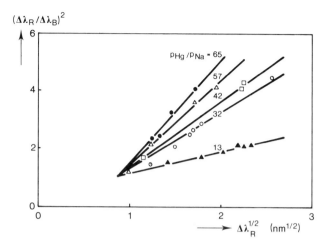

Figure 5.24 The relation between $(\Delta\lambda_R/\Delta\lambda_B)^2$ and $\Delta\lambda_R^{1/2}$ as found in HPS lamps with various ratios between mercury and sodium vapour pressures. $\Delta\lambda_B, \Delta\lambda_R$ = wavelength separation between the centre of the sodium D-lines and the maximum in the blue wing and in the red wing, respectively. (de Groot and van Vliet, 1979)

Up to now, no ultimate comparison has been made between the results of the various methods mentioned. A check with an independent method for the determination of sodium and mercury vapour pressures (e.g. from the filling, using lamps where all the sodium and mercury is evaporated) would be desirable.

5.6 Influence of Amalgam Composition and Temperature on Sodium and Mercury Vapour Pressures

The partial vapour pressures of sodium and mercury above the liquid sodium amalgam are basically determined by the temperature and the composition of the amalgam. In general, the partial vapour pressure p_i of a component i in an alloy of two components i and j is given by the relation

$$p_i = a_i \, p_i^0 = \gamma_i \, x_i \, p_i^0 \qquad (5.8)$$

where a_i = the thermodynamic activity of component i in the alloy
p_i^0 = the saturated vapour pressure of the pure component i
γ_i = the activity coefficient of component i
x_i = the mole fraction of component i in the alloy

For an ideal solution the activity coefficient $\gamma_i = 1$, so the partial vapour pressure of component i is determined by the mole fraction present (Raoult's law). Figure 5.25 gives the partial vapour pressures of sodium and mercury for an amalgam temperature of 973 K as functions of the sodium and mercury

160

mole fractions as compared with the partial vapour pressures calculated according to Raoult's law. The sodium and mercury partial vapour pressures are considerably lower than the values expected according to Raoult's law, since the activity coefficient, for both components, has a value less than unity.

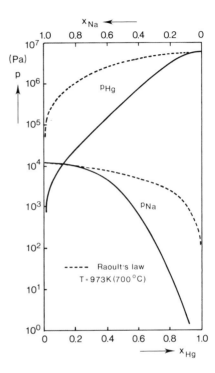

Figure 5.25 Partial vapour pressures p of sodium and mercury (full curves) as functions of the mercury and sodium mole fraction (x_{Hg}, x_{Na}) in the sodium amalgam at a temperature of 973 K (Schmidt, 1968). The calculated vapour pressures according to Raoult's law are also given (dashed curves).

For a long time there has been uncertainty regarding the magnitude of the mercury vapour pressure of sodium amalgams (Wharmby, 1980). On the one hand there were theoretical estimations based on old, unreliable thermodynamic data, and on the other there was a rather limited amount of experimental data given in the patent literature without mentioning the measuring methods employed. It was not until 1981 that the first publication on the measured thermodynamic activities of sodium and mercury in sodium amalgams in the relevant temperature range for HPS lamps appeared (Hirayama *et al.*, 1981a, 1983).

Thermodynamic Data
The value of the thermodynamic activity a_i of component i can be calculated from thermodynamic data (Hultgren *et al.*, 1973)

161

$$\ln a_i = \ln \left(\frac{p_i}{p_i^0}\right) = \frac{\Delta \bar{G}_i}{R_g T} = \frac{\Delta \bar{H}_i - T \Delta \bar{S}_i}{R_g T} \tag{5.9}$$

where R_g = the molar gas constant
$\Delta \bar{G}_i$ = the partial molar free energy of component i
$\Delta \bar{H}_i$ = the partial molar enthalpy of component i
$\Delta \bar{S}_i$ = the partial molar entropy of component i

Assuming that $\Delta \bar{H}_i$ and $\Delta \bar{S}_i$ are temperature independent, it follows that

$$\frac{\partial}{\partial T}(\ln a_i) = -\frac{\Delta \bar{H}_i}{R_g T^2} \tag{5.10}$$

With relation (5.10) and the values for a_i and $\Delta \bar{H}_i$ as given by Hultgren *et al.* (1963, 1973) for temperatures of 648 K and 673 K, the values of a_{Na} and a_{Hg} can be calculated as functions of temperature. From these activity values and the known relations between vapour pressure and temperature for pure sodium and mercury (Stull and Prophet, 1971; Hultgren *et al.*, 1973), the sodium and mercury partial vapour pressures above the amalgam can be calculated from Eq. (5.8). The calculated vapour pressures are given in figure 5.26 as functions of temperature for various amalgam compositions. These calculated values are based on the thermodynamic data given by Hultgren

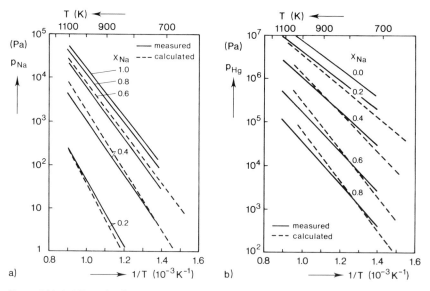

Figure 5.26 (a) Partial sodium vapour pressure p_{Na} and (b) partial mercury vapour pressure p_{Hg} as functions of (inverse) temperature for various values of the sodium mole fraction x_{Na} in the sodium amalgam. The calculated curves are based on the thermodynamic data given by Hultgren et al. (1973); the measured curves are based on the data of Hanneman (1968).

et al. (1973); calculations based on older Hultgren data (1963) give considerably higher values for the mercury vapour pressure.

Vapour Pressure Measurements
The partial vapour pressures of mercury and sodium have been determined experimentally by Hanneman *et al.* from x-ray absorption measurements in the temperature range 715 to 1100 K for amalgams with sodium mole fractions varying from zero to unity. Their data were published, only in part, in the patent literature (Hanneman *et al.*, 1969a; Schmidt, 1968) and are reproduced in figures 5.25 to 5.27. Comparing the results of these measurements with the results of the thermodynamic calculations presented in the previous paragraph (figure 5.26), it appears that the thermodynamic calculations give considerably higher values for the mercury vapour pressure, when considering amalgam temperatures and compositions that may be considered as typical for HPS lamp conditions ($x_{Na} = 0.6 - 0.8$, $T = 900 - 1100$ K). This is because the thermodynamic data employed, which are valid for low-temperature amalgams cannot, as shown by Hirayama *et al.* (1981a), be accurately extrapolated to high-temperature amalgams.

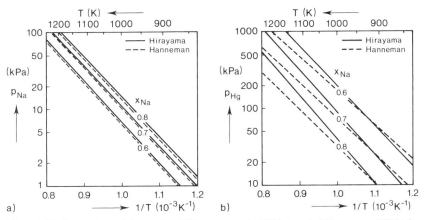

Figure 5.27 Comparison of the results of Hirayama et al. (1981a) and of Hanneman (1968) for (a) partial sodium and (b) partial mercury vapour presssures as functions of the (inverse) temperature. Data are given for various values of the sodium mole fraction x_{Na} in the sodium amalgam, representative for HPS lamp conditions.

The thermodynamic activity of sodium in high-temperature amalgams was recently determined by Hirayama *et al.* (1981a, 1983) from EMF measurements, in the range 771 – 884 K, of electrochemical cells of the type
Na | Na$^+$ conductor | Na(Hg)

The EMF of such a cell is given by the Nernst equation

$$\text{EMF} = -\frac{R_g T \ln a_{\text{Na}}}{F} \tag{5.11}$$

where F is the Faraday constant.

The sodium partial vapour pressures were determined from the measured sodium activities and the known relation between vapour pressure and temperature for pure sodium (Stull and Prophet, 1971). The total pressures p_{tot} above the amalgam were also measured at temperatures of about 800 K to 1000 K. The partial vapour pressures of mercury were obtained by subtracting the sodium partial pressures from the total pressures. These results were summarised by Reiser and Wyner (1985) in the following approximations for mercury vapour pressure p_{Hg} and sodium vapour pressure p_{Na} (in Pa)

$$p_{\text{Hg}} = x_{\text{Hg}} \exp \left[22.4950 - \frac{6865.45}{T} + (15.04924 - \frac{27164.05}{T}) x_{\text{Na}}^2 - (13.5218 - \frac{21948.0}{T}) x_{\text{Na}}^3 \right] \tag{5.12}$$

$$p_{\text{Na}} = p_{\text{Na1}} + p_{\text{Na2}} \tag{5.13}$$

where

$$p_{\text{Na1}} = x_{\text{Na}} \exp \left[21.5643 - \frac{11807.0}{T} - (5.23346 - \frac{5759.30}{T}) x_{\text{Hg}}^2 + (13.5218 - \frac{21948.9}{T}) x_{\text{Hg}}^3 \right] \tag{5.14}$$

and

$$p_{\text{Na2}} = p_{\text{Na1}}^2 \exp \left(-21.2712 + \frac{9631.99}{T} \right) \tag{5.15}$$

In the derivation of these equations it was assumed that the sodium molecules (Na_2) existing at relatively low amalgam temperatures are completely dissociated at the relatively high temperatures in the discharge volume.

In figure 5.27 the partial sodium and mercury vapour pressures are given for typical HPS lamp conditions as calculated with Eqs (5.12) and (5.13). These results, based on the data of Hirayama et al. (1981a), give 5 to 10 per cent higher sodium vapour pressures compared with the data of Hanneman (1968), which are also shown in this figure. For temperatures between 900 K and 1000 K the mercury vapour pressure data of Hirayama et al. are in good agreement with the values of Hanneman. They are higher than the values given by Hanneman for temperatures over 1000 K; at 1200 K the differences may amount to some 40 per cent. At these high temperatures, extrapolated values are the only data available. It would be desirable to have direct measurements of the mercury partial vapour pressure at these high temperatures also. From the data presented, it can be concluded that the ratio of mercury to sodium vapour pressure depends largely on the sodium mole fraction in the amalgam, while the dependence on the amalgam temperature is only weak (see figure 5.28). The mercury partial vapour pressure is higher than that of sodium (because of the higher volatility of mercury) when the sodium mole fraction in the amalgam is lower than 0.85 ($x_{Hg} > 0.15$).

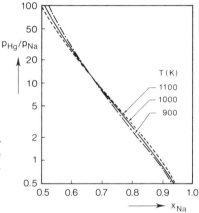

Figure 5.28 The ratio of mercury to sodium vapour pressure as a function of the sodium mole fraction x_{Na} in the amalgam at various temperatures as derived from the data of Hirayama et al. (1981a).

HPS Lamp

The sodium and mercury vapour pressures in HPS lamps are in principle determined by the temperature and the composition of the sodium amalgam at the coldest spot in the discharge volume. However, it is not always easy to translate the relations between partial vapour pressures and temperature and composition of the amalgam, as discussed in the preceding paragraphs, into practical lamp situations. A complication is that in a practical HPS lamp the actual amalgam composition under operating conditions deviates from that of the dosed composition.

Mercury Depletion and Sodium Loss

From the data presented in figure 5.28 it can be concluded that the mercury mole fraction in the vapour is much higher than that in the liquid. This leads to a depletion of mercury in the liquid amalgam, as the amount of dosed amalgam is limited. The change in amalgam composition depends on the dosed amalgam mass and the amount of mercury evaporated, which in turn depends on the amalgam temperature, the discharge tube volume and the average temperature in the discharge tube. Another effect that may change the actual amalgam composition is the sodium loss caused by physical or chemical binding of the sodium (Secs 8.1.4. and 10.1.2).

The actual sodium mole fraction in the amalgam can be calculated, with an iterative procedure, from the following relations

$$x_{Na} = \frac{m_{Na(l)}/M_{Na}}{m_{Na(l)}/M_{Na} + m_{Hg(l)}/M_{Hg}} \tag{5.16}$$

$$x_{Hg} = 1 - x_{Na} \tag{5.17}$$

$$m_{Na(l)} = m_{Na} - m_{Na(g)} - \Delta m_{Na} \tag{5.18a}$$

$$m_{Hg(l)} = m_{Hg} - m_{Hg(g)} \tag{5.18b}$$

$$m_{Na(g)} = \frac{M_{Na}}{N} \int_{vol} \frac{p_{Na}(x_{Na}, T_{am})}{k\,T} \, dV \tag{5.19a}$$

$$m_{Hg(g)} = \frac{M_{Hg}}{N} \int_{vol} \frac{p_{Hg}(x_{Na}, T_{am})}{k\,T} \, dV \tag{5.19b}$$

where
x_{Na}	= sodium mole fraction in liquid amalgam
x_{Hg}	= mercury mole fraction in liquid amalgam
$m_{Na(l)}, m_{Hg(l)}$	= sodium and mercury mass in liquid amalgam
$m_{Na(g)}, m_{Hg(g)}$	= sodium and mercury mass in gas (vapour) phase
m_{Na}, m_{Hg}	= total dosed sodium and mercury mass
Δm_{Na}	= sodium loss
M_{Na}, M_{Hg}	= molar mass of sodium and mercury respectively
N	= Avogadro number
T_{am}	= amalgam temperature
T	= gas temperature in discharge volume
k	= Boltzmann constant
vol, V	= volume of discharge tube

In figure 5.29 the calculated sodium mole fraction in the amalgam during lamp operation is given as a function of the amalgam temperature for various amounts of the amalgam dosed and for an assumed effective gas temperature of 2500 K. From this figure it can be concluded that for practical HPS lamps with amalgam weights per unit of volume of between 3 and 50 μg mm^{-3}, the mercury depletion effect significantly influences the amalgam composition during operation. This has also been observed experimentally by Zollweg and Kussmaul (1983). It also follows from measurements of the amount of mercury near the coldest spot in the discharge tube, as can be done with radiochemical methods. During lamp operation the amount of mercury near the coldest spot may be significantly lower than in case where the discharge tube is cold (figure 5.30). Further, figure 5.30 illustrates that it may be difficult

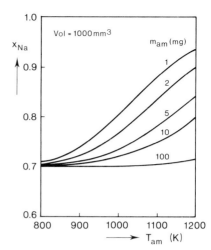

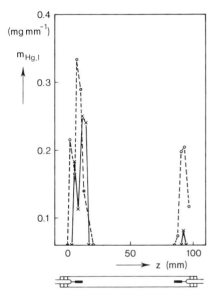

Figure 5.29 Effect of amalgam mass on the sodium mole fraction x_{Na} in the amalgam during lamp operation as a function of amalgam temperature T_{am}. The calculations have been carried out for various amalgam masses for a discharge volume of 1000 mm^3 (1 cm^3) and a dosed sodium mole fraction $x_{Na} = 0.7$, assuming an effective gas temperature $T = 2500$ K.

Figure 5.30 Distribution of mercury in the discharge tube of a 360 W HPS lamp during operation (full curve) as well as in off (cold) situation (dashed curve). A radiochemical method was employed, the detector having a 1 mm slit; $m_{Hg,l}$ is the mass of mercury per unit of length. (Bruijs and Schellen, 1978)

to measure the amalgam temperature in actual HPS lamps, because the amalgam is not always exclusively present at the coldest spot. For the lamp shown in figure 5.30 the amalgam is present over a length of 10 to 20 mm, despite an estimated temperature difference of about 200 K along this length.

Figure 5.31 shows the calculated effect of finite amalgam mass on the relation between mercury vapour pressure and sodium vapour pressure. Because of the mercury depletion effect, the mercury vapour pressure is lower and the sodium vapour pressure higher for a given amalgam temperature, than expected for an infinite amalgam mass. In experimental discharge tubes the amount of amalgam can be made so large that the effect of mercury depletion is negligibly small (Denbigh and Wharmby, 1976). With an appropriate amalgam weight and composition the amalgam is completely in the vapour phase above a certain amalgam temperature, and an unsaturated vapour lamp is obtained.

167

The effect of sodium loss is shown in figure 5.32. This effect causes the amalgam to become richer in mercury, and this increases the mercury vapour pressure for a given amalgam temperature. The data as given in figures 5.31 and 5.32 can be used, in combination with the data presented in figure 5.7, to

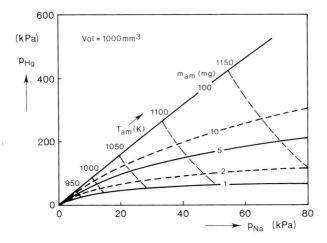

Figure 5.31 Decrease in mercury vapour pressure for given amalgam temperature T_{am} when the amalgam mass in the lamp is lowered. The calculated mercury vapour pressure p_{Hg} in an HPS discharge is given as a function of the sodium vapour pressure p_{Na} for various values of the amalgam mass for a discharge volume of 1000 mm³ (1 cm³). The calculations were performed for a dosed sodium mole fraction $x_{Na} = 0.7$ and they were based on the sodium and mercury vapour pressures as functions of amalgam composition and temperature as given by Eqs (5.12-5.15).

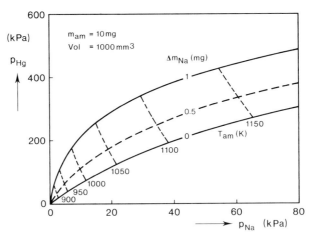

Figure 5.32 Increase in mercury vapour pressure, for given amalgam temperature T_{am}, when there is a loss of sodium. The calculations have been carried out for three values of the sodium loss Δm_{Na} for a dosed amalgam mass of 10 mg per 1000 mm³ (1 cm³).

168

calculate the influence of amalgam weight and sodium loss on the electric field strength (see also Sec. 10.1.2).

Because of the various effects mentioned, it will be clear that it is important to have available reliable methods for measuring the actual sodium and mercury vapour pressures in the practical lamp situation.

5.7 Final Remarks

From the contents of this chapter it will be clear that a buffer gas may play an important role in determining the ultimate properties of the HPS lamp. The buffer gas type and pressure may be chosen to influence the electrical as well as the spectral characteristics. The highest radiant efficiency is found for the combination of the buffer gases mercury and xenon. The present model for the power balance of the HPS lamp gives a satisfactory description of electric field strength and plasma temperature in HPS discharges with various buffer gases, but is not yet adequate to give a full description of the influence of the buffer gas on the radiant power.

Methods are available for determining the sodium pressure as well as that of the buffer gas under actual lamp conditions, making it possible to better understand and describe the discharge properties. However, a more complete evaluation of the various methods remains desirable.

Chapter 6

Ignition and Stabilisation

A characteristic common to all discharge lamps, including the HPS lamp, is the need to ignite and stabilise the discharge.

Ignition involves conversion of the starting gas from a non-conductive state into a conductive state, with the glow discharge eventually developing into an arc discharge. The first important stage in the ignition process, the breakdown of the starting gas (and thus the ignition itself) can only be achieved if the electrical circuit provides the lamp with a starting voltage of sufficient amplitude and appropriate width and rise time.

Stabilisation involves limiting the current flowing through the discharge which, because of the negative voltage-current characteristic of the arc discharge, would otherwise increase indefinitely and destroy the lamp. Thus, the lamp must be operated from a current limiting device. The lamp current can be adjusted to the desired value by placing a passive ballast or an electronic ballast between the lamp and the supply.

To achieve proper ignition and stabilisation certain requirements must be met by the discharge lamp, and by the electrical circuit to which the lamp is connected.

The ignition and stabilisation of HPS gas discharge lamps will be described in Secs 6.1 and 6.2 respectively.

6.1 Ignition

The process of ignition will be illustrated schematically by the phenomena in a discharge between plane-parallel plates. The relevant voltage-current characteristic is shown in figure 6.1. A very small, intermittent current flows through the gap when a certain, relatively low voltage is applied. So-called primary electrons in the gas, or liberated by photo-electric effect or by cosmic radiation from the cathode, move to the anode. In order to increase the average value of the current, the voltage must be raised. In this 'Geiger' region (I) the primary electrons are accelerated in the homogeneous electric field between the electrodes and multiplied by ionisation of the gas atoms. The value of the average current is determined by the number of primary electrons generated per second as well as by the energy the electrons acquire in the

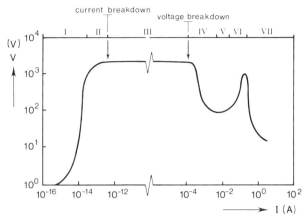

Figure 6.1 Schematic voltage-current characteristic of a gaseous gap between two electrodes. The voltage (V) and current (I) scales are logarithmic. The discharge regions are indicated

I   Geiger region
II   Townsend discharge
III   Self-sustained current discharge
IV   Subnormal glow discharge
V   Normal glow discharge
VI   Abnormal glow discharge
VII Arc discharge.

electric field. In the 'Townsend' region (II) the current is still intermittent, the average current increasing strongly for a very small voltage rise. Then, at the point of 'current breakdown', the discharge becomes self-sustaining (III), viz. each primary electron provides for at least one successor by means of one or other secondary electron-liberating process, e.g. the release of electrons by positive ions impinging on the cathode. The discharge current is no longer intermittent. The voltage changes slightly if, after current breakdown, the current is further increased to the point where 'voltage breakdown' occurs. Here the 'subnormal glow discharge' region (IV) sets in, and a substantial voltage drop takes place. At still higher currents the 'normal glow' discharge region (V) is attained, where the voltage is nearly constant. Then, with further increase of the current the normal glow develops into the 'abnormal glow' discharge (VI), where the voltage rises once more. The transition from the latter region to the 'arc discharge' (VII) with considerably lower voltage is only possible if the cathode is locally heated to such a high temperature that ample thermionic emission sets in.

In short, during the ignition process the discharge passes through a succession of stages, the most important of which are: current breakdown, leading to the self-sustaining discharge; voltage breakdown, leading to the glow discharge; and glow-to-arc transition, leading to the arc discharge.

The breakdown mechanisms in tubular discharge vessels as used for HPS lamps are different from those in a discharge gap with plane-parallel electrodes described so far. Nevertheless, the condition for a self-sustaining discharge is the same, viz. that each primary electron starting from the cathode must ultimately give rise to at least one successor. The phenomena of 'current breakdown' and subsequent 'voltage breakdown' are also found in these vessels. And, again, ample thermionic emission of the cathode is necessary for the transition of the abnormal glow into an arc.

The glow-to-arc transition may occur immediately after voltage breakdown, and possibly also at a number of current reversals after reignition. Whether or not an arc is sustained will depend on the power supplied by the electrical circuit to the discharge, especially to the electrodes. Obviously, for proper ignition, viz. for the early development of the discharge (Sec. 6.1.1), glow-to-arc transition (Sec. 6.1.2) and reignition at current zero (Sec. 6.1.3), the discharge lamp and the electrical circuit to which it is connected should be well matched.

6.1.1 Early Development of the Discharge

The first stages in the development of the discharge will be treated in this section as well as the technical means used to facilitate ignition during these stages. Two current breakdown mechanisms that may be active in HPS lamps will be considered: the Townsend mechanism and the streamer mechanism. Not only the nearly equal voltages for current breakdown and voltage breakdown are of interest for the designer of lamps and circuits, the temporal development of the discharge is equally important. Both breakdown and temporal development are investigated in experimental breakdown studies. More practical points are raised on rise time, ignition devices, starting aids and hot restrike.

Breakdown

During the breakdown phase, current breakdown is followed by voltage breakdown. From a practical point of view voltage breakdown is the most conspicuous stage of the breakdown process, and is easily established experimentally. Voltage breakdown will henceforth be referred to as 'breakdown'; current breakdown will be indicated in full.

The first breakdown stage, current breakdown, leads to a self-sustaining discharge. In a self-sustaining discharge in the inter-electrode space of a discharge lamp, each primary electron liberated at the cathode must give rise to at least one successor, viz. one secondary electron.

From the literature, two current breakdown mechanisms are known: the Townsend and the streamer mechanisms. These mechanisms are discussed in general terms below. A detailed description falls outside the scope of this

172

book, and for full information the reader is referred to the book of Nasser (1971).

Townsend Mechanism

The Townsend mechanism is responsible for current breakdown when the space-charge field at the head of the electron avalanche is too low to distort the applied field. The first electron avalanche causes a series of successive avalanches in the total discharge volume until current breakdown occurs. The secondary electrons may be liberated in one of two ways: by ion bombardment at the cathode, or by photons created between the electrodes.

Current breakdown can be illustrated by considering a discharge between two plane-parallel electrodes, in which it is assumed that the secondary electrons are obtained from bombardment of the cathode by positive ions. In the homogeneous electric field between two plane-parallel electrodes, each primary electron produces $[\exp(\alpha l) - 1]$ ions on its way to the anode, where α (the Townsend electron ionisation coefficient) is the probability per unit length of path in the direction of the electric field that an electron will ionise an atom – thereby producing a new electron and an ion – and l is the effective distance available for ionisation between cathode and anode. As the difference between l and the electrode spacing is small, especially at high gas pressures, this difference will be neglected in the following. Suppose that all these ions collide with the cathode; the well-known Townsend criterion for current breakdown is then fulfilled if

$$\gamma_i[\exp(\alpha l) - 1] = 1 \tag{6.1}$$

where γ_i = secondary electron emission coefficient of the cathode by ion bombardment

α = Townsend (electron) ionisation coefficient

l = discharge length between cathode and anode

The Townsend ionisation coefficient is strongly dependent on the electric field strength and can, for a variety of gases, be described by (Nasser, 1971)

$$\alpha/p_o = A \exp(-Bp_o/E) \tag{6.2}$$

where p_o = gas pressure at temperature $T = 273$ K

E = electric field strength

A and B are constants depending on the type of gas used

Combining Eqs (6.1) and (6.2) leads to the following relation for the current breakdown voltage V_b in an homogeneous space-charge-free electric field

$$V_b = El = \cfrac{Bp_o\, l}{\ln\left[\cfrac{Ap_o\, l}{\ln(1 + \cfrac{1}{\gamma_i})}\right]} \tag{6.3a}$$

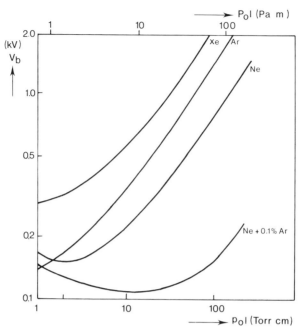

Figure 6.2 Measured Paschen curves for some noble gases and a neon/argon Penning mixture. p_o is the gas pressure at 273 K and l is the distance between the electrodes. The measurements in xenon were done for parallel electrodes made of nickel (Bhattacharya, 1976); the measurements in neon, argon and neon/argon mixture were for plane parallel electrodes made of molybdenum. (Frouws, 1957)

The current breakdown voltage is generally a function of the product of reduced gas pressure and electrode spacing, $p_o l$. Therefore, the results of breakdown voltage measurements are generally presented in so-called Paschen curves giving V_b as functions of $p_o l$. Such curves for some noble gases as well as for the mixture of 99.9 per cent of neon and 0.1 per cent of argon are shown in figure 6.2. The breakdown voltage decreases with increasing $p_o l$ at small $p_o l$ values and increases with increasing $p_o l$ at large $p_o l$ values, so that there exists a minimum breakdown voltage for one distinct value of $p_o l$. If γ_i does not depend on $p_o l$, i.e. on E/p_o, and if α/p_o satisfies Eq. (6.2), the value of $p_o l$ for which the minimum of V_b occurs can be found from Eq. (6.3). This value is

$$(p_o l)_{min} = \frac{2.72}{A} \ln \left(1 + \frac{1}{\gamma_i}\right) \tag{6.3b}$$

As to the values of V_b at a given value of $p_o l$ for the various noble gases, figure 6.2 shows that V_b is highest for xenon. It is followed by argon and

174

neon, whereas the mixture of neon and argon exhibits by far the lowest value of V_b. With such a so-called Penning* mixture, the neon atoms (main gas) are excited to give long-life metastable atoms at an energy level of 16.6 eV. These metastable atoms are able to ionise the argon atoms (additional gas) the ionisation energy of which is 15.7 eV, or just slightly lower than the 16.6 eV neon level.

In tubular discharge vessels and for the pressures used in HPS lamps, the current breakdown voltage V_b increases with increasing pressure as indicated in the Paschen curves, and the sequence of the noble gases and their mixtures with respect to decreasing V_b values is also the same as with the curves shown in figure 6.2.

Most of the commercially available HPS lamps use xenon as the starting gas. If looked at solely from the point of view of breakdown, this is the worst possible gas to choose for a starting gas, especially since in practice the pressures are substantially higher than the one corresponding to the minimum of V_b in the Paschen curve†. The choice is made because xenon has a low thermal conductivity, which leads to a high luminous efficacy. The drawback of a higher current breakdown voltage is accepted. The best starting would, in fact, be obtained using a Penning mixture consisting of 99.9 per cent neon and 0.1 per cent of argon. However, as compared to xenon, the use of such a neon-argon gas mixture suffers from two drawbacks. In the first place, the relatively high thermal conductivity of neon decreases the luminous efficacy of the lamp (Sec. 5.3). Secondly, such a mixture of light atoms at a low pressure causes sputtering at the cathode during a relatively long period after breakdown, and this shortens lamp life. It is only in those HPS lamps that are designed as direct replacements for high-pressure mercury lamps in existing installations that these Penning mixtures are used – in combination with an external conductor as a starting aid (see figure 6.11b) – the object being to give these lamps a breakdown voltage below the periodic maximum of the supply voltage (Cohen and Richardson, 1975; Collins and McVey, 1975b; de Neve, 1976).

Streamer Mechanism

If the gas pressure is high and the amplitude of the starting voltage greatly exceeds the minimum voltage needed for breakdown, then another break-

---

* F. M. Penning (1929) was the first to investigate this effect of small admixtures thoroughly and to explain its mechanism.

† A feature of the ignition process that will not be discussed here is the trouble that can be caused in some new lamps by the presence of pollution gas, principally hydrogen. When a new lamp is operated for the first time, the hydrogen rapidly disappears from the discharge tube via the 'hydrogen window' – the niobium at the ends of the discharge tube (Sec. 8.2) – to be absorbed by the barium getter in the outer envelope.

down mechanism, the so-called streamer mechanism, can be used to explain current breakdown. In this mechanism a channel of conductivity is rapidly developed between the electrodes during the time that one avalanche is building up. This mechanism is based on a local field enhancement by space charges in the region of the head of the avalanche. It is generally assumed that the streamer sets in when the space-charge field becomes of the same magnitude as the applied field. The streamer mechanism explains, among other things, the branched filamentary breakdown channel visible when HPS lamps with high xenon starting gas pressures are ignited (Zakharov *et al.*, 1974).

The second breakdown step is voltage breakdown. It has already been remarked that the discharge voltage at this step is at most slightly higher than the current breakdown voltage V_b. Voltage breakdown can easily be established experimentally because of the substantial discharge voltage drop following this step.

Temporal Development
In the breakdown process, the following stages can be discerned (figure 6.3)
a) After a certain time $t_{b,I}$, current breakdown occurs. This is the so-called formative time lag needed to build up an electron avalanche so that new electrons can be generated by its by-products (i.e. ions, metastable atoms or photons) to get a self-sustaining discharge. The statistical time lag, i.e. the average time elapsing between the application of the voltage and the

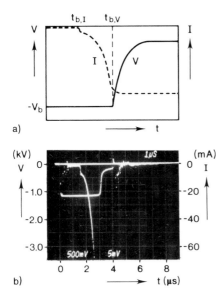

Figure 6.3 Temporal development of current and voltage during the breakdown process (a) schematic, (b) oscillogram. At the moment of current breakdown ($t_{b,I}$) a conductive channel has been formed between the electrodes. At the moment of voltage breakdown ($t_{b,V}$) a collapse of the voltage occurs, and a glow or arc discharge is initiated. The current breakdown can be clearly seen in the oscillogram for a 400 W HPS lamp. The current breakdown only becomes clearly visible in the oscillograms if the sodium vapour pressure during starting is elevated to a few Pascal. (The coldest spot temperature equals 500 to 600 K.)

176

appearance of a primary electron to initiate an electron avalanche, can generally be neglected as compared to the total time needed for the breakdown process. This is due to the rather large volume of the discharge tube and to the low work function of sodium and of the emitter used on the electrodes.

b) After a time $t_{b,V}(> t_{b,I})$, when the current is large enough for the subnormal glow discharge to set in, voltage breakdown occurs.

c) After breakdown ($t > t_{b,V}$), either a glow discharge or an arc discharge will take place, depending on the amount of energy fed into the cathode after breakdown. This stage will be treated in Sec. 6.1.2.

The formative time lag depends on which mechanism is responsible for current breakdown. If this is the Townsend mechanism, and if secondary electrons are produced by ion bombardment at the cathode, the formative time lag will always be larger than the time needed for an ion to reach the cathode from the place where it is produced. The minimum path an electron has to cover on average to ionise an atom is given by the reciprocal of the Townsend ionisation coefficient (Nasser, 1971). The minimum formative time lag t_{min} is therefore given by

$$t_{min} = \frac{1}{\alpha v_i} \tag{6.4}$$

where v_i is the ion drift velocity.

When secondary electrons are produced by ion bombardment t_{min} is of the order of milliseconds. If the secondary electrons are produced by photons created between the electrodes, then in Eq. (6.4) the ion drift velocity has to be replaced by the electron drift velocity, and the formative time lag is determined by the time needed by an electron to cover the path to ionise an atom. In this case the value of t_{min} is in the order of microseconds.

The development of a streamer into a conductive channel takes place so rapidly that the formative time for streamer breakdown is primarily determined by the time required for the space-charge formation, and is in the order of microseconds, or even smaller.

The moment of current breakdown in HPS lamps normally occurs within a time of 0.1 μs. However, at a temperature of about 500 K this moment may be delayed by several microseconds (figure 6.3). It is assumed that both the Townsend and the streamer mechanisms may be active in HPS lamps. For relatively low voltages above the minimum required for breakdown, the formative time lag is principally defined by the Townsend mechanism, while for high voltages it is the streamer mechanism that defines $t_{b,I}$.

After a time $t_{b,V}$ voltage breakdown occurs. The greater part of this time is needed to increase the conductivity of the breakdown channel to the point where the current is large enough for the subnormal glow discharge to set in. Generally, the time for breakdown is much larger than the formative time lag.

Experimental Breakdown Studies

The breakdown process is studied by supplying the non-conducting HPS lamp with a constant voltage whose amplitude can be chosen (Collins and Wenner, 1976). Breakdown is then observed as a sudden voltage drop (figure 6.4).

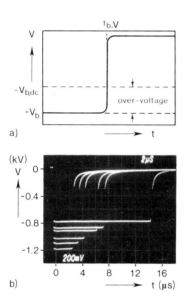

Figure 6.4 Influence of the magnitude of the over-voltage on the time for breakdown (a) schematic, (b) oscillogram. Voltage break-down occurs $t_{b,V}$ seconds after supplying the HPS lamp with a constant voltage larger than the minimum voltage ($V_{b,dc}$) required for breakdown. The larger this over-voltage, the shorter the time $t_{b,V}$. The oscillogram shows this behaviour for six values of the supplied voltage. (van Vliet, 1983)

To obtain reproducible measurements of the breakdown voltage and of the time to breakdown, HPS lamps must first be operated for at least ten minutes on rated current, followed by a cooling-off period of the same duration. Inadequate preliminary burning or cooling-off will influence the results of the measurements. This is illustrated in figure 6.5, where the measured breakdown voltage for $t_{b,V} = 1$ μs is shown as a function of the cooling-off time. For short cooling-off times the breakdown voltage becomes very high because of the high sodium and mercury vapour pressures present. A minimum in the breakdown voltage is found for a cooling-off time of between one and two minutes. This minimum may be explained by the possible presence of a Penning mixture of xenon and sodium after such a period of cooling. For cooling times larger than about five minutes, the breakdown voltage is mainly determined by the presence of the xenon starting gas.

178

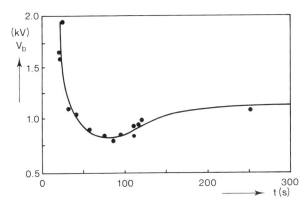

Figure 6.5 Influence of the cooling-off time on the breakdown voltage for a 70 W HPS lamp $(t_{b,V} = 1 \mu s)$.

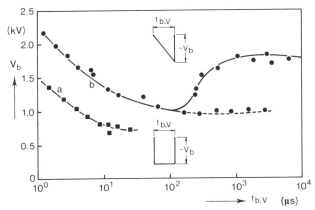

Figure 6.6 Relation between time to breakdown and breakdown voltage of a 70 W HPS lamp for (a) a constant supply voltage and (b) a voltage increasing linearly with time. In the latter case, two amplitudes for the breakdown voltage were measured between 200 μs and a few milliseconds. (van Vliet, 1983)

Operating periods shorter than about two minutes must also be avoided, or the breakdown voltage may be increased. This increase is possibly caused by the changed condition of the surface of the electrode and of the wall next to the electrode.

If the amplitude of the voltage applied in breakdown experiments is greater than the minimum required for breakdown, the time to breakdown will largely depend on the so-called overvoltage (figure 6.4), which is the difference between the applied voltage and the minimum required for breakdown. This minimum breakdown voltage $V_{b,dc}$ is found if the breakdown process is not limited in time. The relation between the breakdown voltage and the time for breakdown is given in figure 6.6 (curve a) for a 70 W HPS lamp. This curve defines the minimum amplitude and width of the rectangular ignition or starting pulse needed for breakdown.

Rise Time

In reality, the starting pulse generated by the ignitor is not rectangular in shape as was assumed in the foregoing. It is necessary, therefore, to examine the influence that the pulse shape has on the magnitude of the breakdown voltage. This can only be done in broad lines. An important feature appears to be the time to breakdown, viz. the time $t_{b,V}$ between the very outset of the ignition pulse and the moment of breakdown. In order to find the influence of this time on the breakdown voltage V_b in a well-defined situation, the discharge lamp is provided with voltages with various rise times that increase linearly with time.

In the oscillogram shown in figure 6.7, the lamp voltages are given as functions of time for eight rise-time values. For small rise times the breakdown voltage decreases with increasing $t_{b,V}$, as is also shown in figure 6.6 (curve b). For large rise times, however, ($t_{b,V} > 0.6$ ms) the breakdown voltage suddenly increases. Between 200 μs and a few milliseconds two amplitudes for the breakdown voltage are found. Such an increase in breakdown voltage at large rise-time values occurs mainly if the electrode spacing is so large with respect to the diameter of the discharge tube that the discharge tube wall is involved in the breakdown process. It could be that within a few hundred microseconds a field has been built up by the ions that opposes the external field, so causing the higher breakdown voltage.

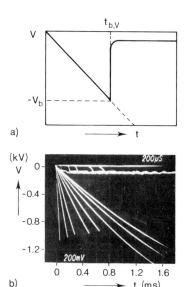

a)

b)

Figure 6.7 Influence of the rise time of the ignition pulse on the magnitude of the breakdown voltage for a 70 W HPS lamp: (a) schematic, (b) oscillogram. The moment of voltage breakdown occurs $t_{b,V}$ seconds after applying a linearly increasing voltage. The oscillogam shows this behaviour for eight rise times of the supply voltage. If the rise time is made small ($t_{b,V} < 0.6$ ms), then the required breakdown voltage decreases with increasing rise time. For large rise times ($t_{b,V} > 0.6$ ms) the breakdown voltage suddenly increases. (van Vliet, 1983)

Ignition Devices

HPS lamps employing xenon as a starting gas are operated in conjunction with an ignition device which, in combination with the inductive ballast or a separate pulse transformer, produces the high voltage needed for breakdown in the form of starting pulses superimposed on the supply voltage. These pulses, shown in figures 6.8 and 6.9 for an electronic and for a glow-switch starter respectively, are not rectangular in shape. With such non-ideal starter pulses, breakdown is only assured if the pulse contour falls outside that of a rectangular pulse that would produce breakdown, and if the rise time is not too long. These starting pulses have therefore to satisfy certain minimum requirements with regard to their amplitude, width and rise time. Their position relative to the peak of the supply voltage is also important. On the other hand, the open-circuit characteristics of the starting pulses must also be considered. Amongst other things, flashover between the supply leads in the base of the lamp is to be avoided.

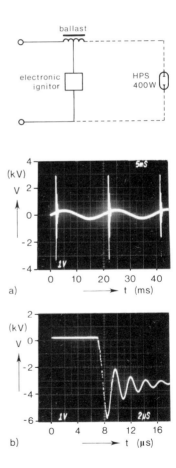

Figure 6.8 Starting pulses, generated by an electronic semi-parallel starter circuit, represented in two time scales. As can be seen from oscillogram (a), the pulse comes slightly before the maximum in the mains voltage.

181

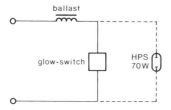

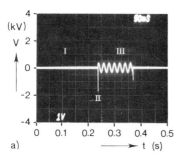

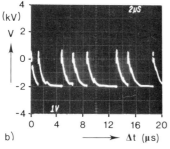

a)

b)

Figure 6.9 Starter pulses, generated by a glow-switch starter represented in two time scales. In diagram (a) three situations of the starter can be discerned: (I) after the bimetals have been heated by a glow discharge, they close and the starter voltage becomes zero; (II) the bimetals are repeatedly closing and disconnecting; the (abnormal) glow discharge in the starter defines the maximum amplitude of the ignition pulse (see b); (III) the bimetals open and are again heated by a glow discharge until they close again (situation I). The glow voltage during the heating-up phase is slightly lower than the mains voltage. V is the voltage over the glow-switch.

An ignition device can be employed in the lamp circuit in the three ways shown in figure 6.10 (Williams, 1970). Figure 6.10a depicts a so-called parallel starter circuit. A switch, which is generally an integral part of the lamp, repeatedly opens and closes to generate voltage transients across the electrodes of the discharge tube, the transients arising from the rapid current changes in the choke. The switch can comprise

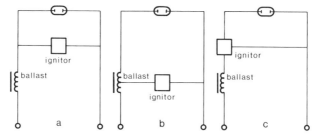

Figure 6.10 Schematic representation of three different ignition circuits: (a) parallel circuit; (b) semi-parallel circuit and (c) series circuit.

182

a) A glow-switch starter consisting of a gas-filled bulb containing a bimetal switch (colour plate 6a). The amplitude of the voltage transient produced during the opening of the switch is limited by the voltage of the (abnormal) glow discharge in the switch itself (figure 6.9). Another bimetal switch in series with the glow-switch starter puts the glow-switch out of circuit at high temperatures. This prevents the discharge being short-circuited by the glow-switch at the high operating temperatures or at reignition voltages higher than the breakdown voltage of the glow-switch.

b) A bimetal switch, sometimes in series with a resistance. The amplitude of the voltage transient produced by the opening of the switch is then limited by the series resistance. This prevents the voltage transient from becoming too high in amplitude, should the lamp fail.

The ignition devices employed in the semi-parallel and series circuits of figure 6.10 generally contain a semiconductor switch. Such ignition devices are classified as electronic ignitors. The closing of the switch causes a charged capacitor to discharge into a few windings of the choke (semi-parallel circuit) or into the primary of the transformer of the ignitor (series circuit). The voltage across these windings is stepped up to the required amplitude to appear across the secondary windings.

The advantage of the series starter circuit with the pulse transformer is that the inductive ballast, that is to say the choke, is not subjected to the high-amplitude voltage transients. Ballasts of high dielectric strength are needed if the ballast has to withstand the ignition voltages generated. However, a disadvantage of the pulse transformer is that it has to be located close to the lamp – to limit the capacitive load – for good performance.

Starting Aids

For HPS lamps containing xenon at high pressure as a buffer gas, ignition with the standard pulse used in starting tests (IEC, 1979) is not possible unless a starting aid is employed. For HPS lamps containing a Penning mixture as the starting gas a starting aid is also used to ignite the lamp on the available mains voltage. The function of the starting aid is to reduce the effective electrode separation so that the $p_o l$-value decreases and lies nearer to the minimum in the Paschen curve. The Paschen curves for HPS discharge tube configurations will, of course, deviate from those valid for closely-spaced flat-plate electrodes where the electric field is approximately homogeneous. The tendency shown, however, will be the same.

Some of the starting aids used in conjunction with HPS lamps are depicted schematically in figure 6.11. Figure 6.11a shows a starting aid in the form of an ignition wire very close to the outer wall of the discharge tube. This wire is connected to one of the electrodes via a capacitor, which presents a short-circuit to the high-frequency starter pulses but a high impedance for the mains frequency. A 150 W HPS lamp equipped with such an aid can be started reliably on a standard ignition pulse of 2800 V amplitude and 2 μs width at xenon pressures up to 25 kPa, as is shown in figure 6.12.

HPS lamps designed as direct replacements for high-pressure mercury vapour (HPMV) lamps contain a Ne/Ar Penning mixture to facilitate starting, plus an ignition aid in the form of a tungsten wire coiled round the discharge tube. This coil is connected to one of the electrodes via a thermal switch (figure 6.11b) and is placed at such a position between the electrodes that the lowest breakdown voltage is achieved. The thermal switch opens once the lamp is warm, and so avoids creating large voltage gradients across the wall of the discharge tube.

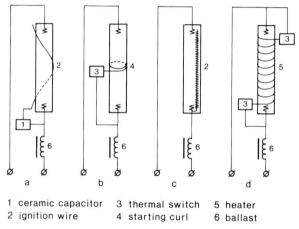

1 ceramic capacitor   3 thermal switch   5 heater
2 ignition wire   4 starting curl   6 ballast

Figure 6.11 Four possible starting aids for HPS lamps: (a) Ignition wire very close to the discharge tube; (b) Ignition wire in the form of a curl round the discharge tube; (c) Ignition wire made of tungsten along the inside wall of the discharge tube; (d) Ignition wire, which serves primarily as a heater to decrease the breakdown voltage, wrapped round the discharge tube.

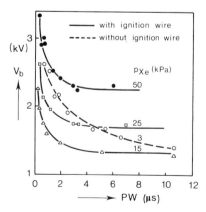

Figure 6.12 Influence of width of starting pulse on breakdown voltage, with and without an ignition wire, for an 150 W HPS lamp, at various xenon starting gas pressures. (van Vliet and Jacobs, 1980).

184

HPS lamps of the starterless variety with an ignition device consisting of a tungsten wire running along the inside wall of the discharge tube, as shown in figure 6.11c, have been described by Nguyen-Dat and Bensoussan (1977). At room temperature the resistance of the wire, one end of which is connected to one of the main electrodes, has to be low enough to allow the breakdown current to flow, but not so low – at the operating temperature – that the arc is short-circuited.

The breakdown voltage can also be reduced by using an external heater to warm the discharge tube (Cohen *et al.*, 1974). This reduction may be explained by the presence of a Penning mixture at a temperature of about 600 K (see figure 6.13). The wire used for this purpose also serves as an auxiliary starting aid (figure 6.11d).

The starting aids depicted in figures 6.11a and 6.11b are those at present used in production lamps. Types c and d are very difficult to realise in production lamps because of the technological problems involved. This is also true for the internal starting electrodes as used in HPMV lamps.

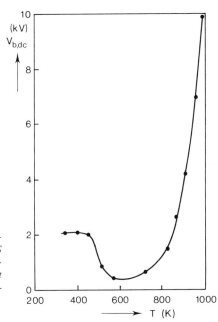

Figure 6.13 Hot-restrike of HPS lamps. Measured d.c. breakdown voltage of a 70 W HPS lamp as a function of the coldest spot temperature of the discharge tube. The minimum at about 600 K is due to the sodium and xenon mixture functioning as a Penning mixture.

Hot Restrike

To restrike a hot HPS lamp a very high amplitude ignition pulse is needed. This is because of the high sodium and mercury vapour pressures present. Without special ignitors, restarting is only possible after a cooling period of half a minute or more (figure 6.5).

In figure 6.13 the d.c. breakdown voltage for a 70 W HPS lamp is given as a function of the temperature of the discharge tube. At a coldest-spot temperature of about 1000 K, ignition pulses with amplitudes above 10 kV are necessary for breakdown. Special HPS lamps have been constructed to withstand these high-voltage pulses so that no breakdown occurs outside the discharge tube; for example, the double-ended lamp, which has an electrical supply lead at each end.

6.1.2 Glow-to-Arc Transition

To complete the ignition process, the electrical circuit should not only initiate voltage breakdown but also provide enough power to the discharge lamp to make the glow-to-arc transition possible. This transition is demonstrated in the oscillogram in figure 6.14a, which shows the lamp voltage immediately after breakdown for various lamp current values. From such measurements the voltage-current characteristics shown in figure 6.14b can be derived. The lamp voltage can have only two distinct values, one belonging to the glow discharge and one to the arc discharge. The arc discharge can only be sustained if the power supplied by the arc to the cathode spot can compensate for the power drain so that, at least locally, the emission temperature is main-

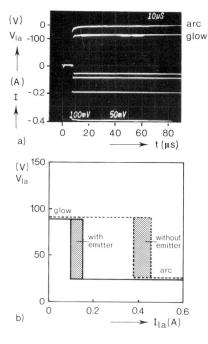

Figure 6.14 Influence of the instantaneous lamp current on the glow-to-arc transition. In the oscillogram (a) the lamp voltage immediately after breakdown is given as a function of time for four lamp currents. The lamp voltage can have only two distinct values, those belonging to the glow and arc discharge phases respectively. From such measurements the voltage-current characteristic of a 70 W HPS lamp immediately after breakdown has been derived (b) for electrodes with and without emitter material. (van Vliet, 1983)

186

tained (Cheng, 1970). The time-dependent power drain is governed in the main by the heat conduction in the cathode material, and thus by the thermal conductivity, the specific heat capacity, and the mass density of the cathode material. The power supplied by the arc is principally determined by the product of lamp current and cathode fall (Sec. 9.3). The cathode fall depends primarily on the work function of the cathode material and on the pressure and composition of the starting gas.

Immediately after breakdown, the electrode dimensions do not play a significant role. Because the power supplied to the cathode surface can be transported into the cathode material only a few micrometres below the surface during the time in which an arc is formed (microseconds), the average temperature of the electrode is approximately equal to the room temperature.

The minimum current at which the transition from glow to arc discharge takes place will be more or less the same for all HPS lamps having the same emitter-coated electrodes and the same starting gas. As shown by the voltage-current characteristic of a 70 W HPS lamp immediately after breakdown (figure 6.14b), below a minimum lamp current ($\approx$ 0.1 A) no arc can exist, while above a maximum lamp current ($\approx$ 0.15 A) no glow can exist. Electrodes not covered with an emitter material need higher emission temperatures and thus larger currents to sustain the arc (figure 6.14b). If, instead of the xenon gas, a Ne/Ar Penning mixture is used as the starting gas, both the glow voltage and the current needed to sustain an arc will be increased.

6.1.3 Reignition at Current-Zero

During the transition to the glow or arc discharge immediately after breakdown, the ignition process can still be hampered at current-reversal. This is caused by the fact that after current reversal the electrical circuit cannot provide the current necessary to sustain an arc at the high glow voltage then prevailing. As long as the electrodes are not heated sufficiently for thermionic emission to take place, these glow phenomena will occur during a fraction of a cycle of the supply voltage. The time, after breakdown, during which such glow phenomena within the cycle occur, is called the ignition time.

In the following sub-section attention will be paid to the influence of the electrical circuit on the reignition behaviour, and to the ignition time.

Influence of Electrical Circuit

For the idealised inductive stabilisation circuit of figure 6.15a, the lamp current can be calculated analytically as a function of time for a given lamp voltage, mains voltage and inductance of the current-stabilising coil. The lamp in such a circuit is assumed to behave like a switch; the voltage across the lamp is assumed to be constant and to change polarity at every current reversal. The current-stabilising coil is free of losses. From the electrical circuit equation

$$\hat{V}_m \sin \omega t = L\frac{di}{dt} + V_{la} \tag{6.5}$$

the current flowing through the lamp immediately after breakdown can now be calculated (Dorgelo, 1937)

$$i(t) = \frac{\hat{V}_m}{\omega L}\left[\cos \omega t_1 - \cos \omega t + \frac{V_{la}}{V_m}(\omega t_1 - \omega t)\right] \tag{6.6}$$

where t = time after the first mains voltage reversal
t_1 = instant of voltage breakdown
L = inductance of the current-stabilising coil
ω = mains angular frequency $(2\pi f)$
$\hat{V}_m$ = amplitude of the mains voltage
V_{la} = lamp voltage

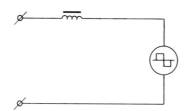

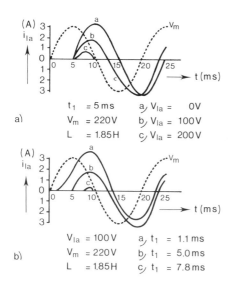

a)

t_1	= 5 ms	a V_{la} =	0 V
V_m	= 220 V	b V_{la} =	100 V
L	= 1.85 H	c V_{la} =	200 V

b)

V_{la}	= 100 V	a t_1	= 1.1 ms
V_m	= 220 V	b t_1	= 5.0 ms
L	= 1.85 H	c t_1	= 7.8 ms

Figure 6.15 Current flow in an idealised lamp circuit immediately after lamp breakdown showing the influence of (a) lamp voltage and (b) time to breakdown. The lamp voltage is assumed to be constant and changes polarity at every lamp-current reversal. The coil is free of losses. The dashed lines give the mains voltage. In (a) breakdown occurs 5.0 ms after the first mains voltage zero. Lamp current waveforms (solid lines) are given for three lamp voltage values. In (b) breakdown occurs 1.1 ms, 5.0 ms and 7.8 ms after the mains voltage passes through zero. (van Vliet, 1983)

188

Thus, for a given mains frequency, the lamp current after breakdown is determined by

a) The ratio of the lamp voltage to the mains voltage.
b) The phase angle $\varphi = \omega t_1$ between the first mains voltage reversal and the occurrence of breakdown.
c) The mains voltage amplitude and the inductance of the ballast.

The lamp current calculated as a function of time after voltage breakdown is given in figure 6.15 for various lamp voltages and phase angles for a mains voltage of 220 V, and a coil inductance of 1.85 H. If the breakdown occurs at the top of the mains voltage, reignition fails when the lamp voltage becomes higher than the reignition voltage available at the first current zero. This is illustrated in figure 6.15a. In the case of an ideal circuit, the reignition voltage available equals the instantaneous mains voltage at current zero. If the moment of breakdown is shifted toward the zero of the mains, a higher reignition voltage is available at the first current reversal (figure 6.15b).

The reignition voltage available is given in figure 6.16 as a function of the time after the mains voltage zero at which breakdown occurs, for several values of the lamp voltage. From this figure it can be concluded that the maximum lamp voltage for which reignition is still possible is about 165 V,

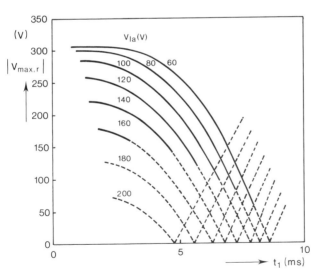

Figure 6.16 Calculated maximum available reignition voltage as a function of the time from the supply voltage zero to where breakdown occurs for several values of the lamp voltage and for a mains voltage of 220 V. The full lines indicate the t_1 and V_{la} values for which reignition at the first current zero is possible. Reignition at first current zero is impossible if the values for t_1 and V_{la} fall on the dashed lines. The maximum lamp voltage at which reignition is still possible is about 165 V.

and this requires that breakdown is achieved not at the top but about 2 ms before the top of the mains.

For practical lamps and circuits the situation is more complicated.

Firstly, the lamp voltage does not have the idealised block waveform. A reignition voltage peak may occur (especially at large current-zero periods) that considerably exceeds the lamp voltage for the case where the current-zero periods are absent. If reignition at current reversal fails, it may be helpful to provide the discharge lamp with ignition pulses in every cycle of the mains. The reignition voltage will then be decreased because the electrodes will be heated and sodium will evaporate, so ensuring that reignition after a number of current reversals will take place.

Secondly, for practical circuits, the mains voltage at current-zero is not instantaneously available for reignition. Therefore, the slower the reignition voltage after current zero is built up to the momentary mains voltage, the

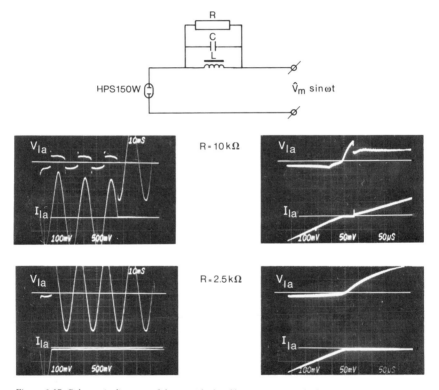

Figure 6.17 Schematic diagram of the non-idealised lamp circuit in which C is the parasitic capacitance of the choke and R its resistance. The oscillograms show the lamp voltage and lamp current during the ignition phase of a 150 W HPS lamp for two time scales. If the lamp voltage after current zero is built-up slowly, as is the case if the resistance R is small, then the current-zero period increases. For too-small resistances reignition fails, and the lamp voltage becomes the mains voltage. (van Vliet, 1983)

190

longer are the current-zero periods. These reignition problems can be demonstrated by bridging the stabilisation coil by a resistor or a capacitor (figure 6.17). The oscillograms of figure 6.17 show the lamp voltage and current during the ignition process for two resistance values. The smaller the resistance, the longer the time before reignition takes place, until the point is reached where reignition fails.

Ignition Time
Glow phenomena may occur during several periods of the mains after breakdown. By way of example, the voltage waveform after breakdown is shown in the oscillograms of figure 6.18 for a 70 W HPS lamp. The time after breakdown during which glow discharges within the cycle are present – the ignition time – is a function of lamp (starting gas, electrodes) and electrical circuit (current, supply voltage) parameters. In figure 6.19 the ignition time is given as a function of the xenon pressure for several values of the mains voltage. From this figure it can be concluded that the higher the mains voltage, the shorter will be the ignition time; the electrodes come up to the emission temperature faster as the power (current) fed to them is increased. That the xenon pressure also has an influence on the ignition time, can be explained from the fact that for a glow discharge the current density increases with the square of the pressure at a constant cathode fall (Engel and Steenbeck, 1934). The power density at the electrodes will thus be enhanced by increasing the pres-

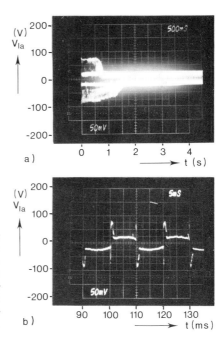

Figure 6.18 Glow phenomena occurring during several periods of the lamp voltage after breakdown of a 70 W HPS lamp stabilised on an inductive ballast at a mains voltage of 198 V. (b) Before and after current zero, glow discharges occur for brief periods within the mains period. (a) After breakdown, some time elapses before an arc occupies both halves of the 50 Hz period.

191

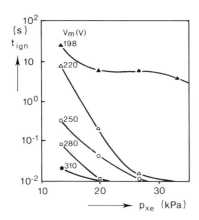

Figure 6.19 Influence of the xenon starting gas pressure on the ignition time of a 50 W HPS lamp for several values of the mains voltage.

sure, from which it can be assumed the electrodes will be more effectively heated at higher gas pressures.

Excessively long glow times have to be avoided, as the sputtering of emitter material during the glow discharge results in blackening of the wall of the discharge tube. This blackening causes an increase in lamp voltage and a gradual depreciation in luminous flux (Sec. 10.1.2).

6.2 Stabilisation

The most important function of the ballast is that of controlling the lamp current. Since the ballast acts as a kind of interface between the discharge lamp and the mains, it must meet the requirements of both the discharge lamp and the mains distribution network.

With respect to the discharge lamp, the ballast must ensure that

– the lamp power is maintained near the rated value, with (small) mains-voltage fluctuations and lamp-voltage variations during life;
– reliable reignition takes place every time the lamp current changes polarity (see also Sec. 6.1.3);
– the proper conditions for breakdown and glow-to-arc transition exist (see also Secs 6.1.1 and 6.1.2).

With respect to the distribution network, the ballast must ensure that

– supply-current distortion is kept within the specified limits needed to maintain an undistorted sinusoidal supply voltage for every user of the supply network;
– in the case of three-phase supplies, high-amplitude third harmonic components in the neutral line are avoided;
– the supply current is as nearly as possible in-phase with the supply voltage so as to keep the cable losses to a minimum; this means that the power

factor, which equals (for small supply current distortion) the 'cos φ' value, should preferably be greater than 0.85;
– radio interference, caused for example during ignition and reignition, is adequately suppressed;
– the ballast does not disturb remote-control signals as, for example, may be used for switching street lighting.

The above requirements with respect to the distribution network will not be dealt with in this book. For a treatment of this topic the reader is referred to the book of Elenbaas (1965).

Additional requirements are, amongst others, that the ballast should dissipate as little power as possible and be as small as possible. It stands to reason, therefore, that the lamp voltage should be kept as high as possible in order to keep the volt-ampere product of the ballast, which more or less determines the ballast losses, as low as possible. Furthermore, it should be noted that the discharge tube of the HPS lamp normally contains an excess of sodium amalgam. During operation, only a small fraction of the amalgam is evaporated. Because of the excess of amalgam, the lamp power influences the coldest spot temperature and the partial pressures of sodium and mercury, and consequently the lamp voltage. As distinct from other types of discharge lamps, such as the HPMV lamp, the lamp power strongly influences the lamp voltage of the HPS discharge. This is illustrated in figure 6.20a.

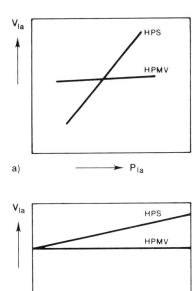

Figure 6.20 Schematic diagrams showing the influence of (a) lamp power and (b) burning time on the lamp voltage for HPS and high-pressure mercury vapour (HPMV) discharge lamps. As distinct from the HPMV discharge lamp, the HPS lamp shows:
a) a considerable increase of the lamp voltage with increasing lamp power, and
b) an appreciable voltage rise during lamp life.

193

Another characteristic of the HPS discharge lamp is the lamp voltage rise during lamp life. This behaviour, too, is coupled with the excess of amalgam present in the discharge tube. As is illustrated in figure 6.20b, the HPMV discharge lamp, all the mercury in which is vaporised during operation, does not exhibit an appreciable rise. However, HPS lamps where all the sodium and mercury are vaporised during operation, are also under development (Hida, 1979).

The excess of amalgam places higher demands on HPS systems than on HPMV systems, both on the discharge lamp itself and on the stabilisation circuit needed to maintain a virtually constant power in the HPS discharge lamp.

The properties of a given combination of ballast and lamp can be read from the so-called ballast and lamp lines.

6.2.1 Ballast Lines and Lamp Lines
As already mentioned, one of the main functions of the ballast is to ensure that, even if there are lamp voltage or mains voltage fluctuations, the correct amount of power is dissipated in the lamp. The way in which lamp power varies with lamp voltage for a given electrical circuit, e.g. a given ballast and a given mains voltage, is shown by the so-called ballast lines. The influence of the lamp voltage on the lamp power during variations in the mains voltage with a given ballast can be illustrated by the so-called lamp lines. The operating point of the HPS discharge lamp is given by the point of intersection of such ballast and lamp lines (figure 6.21).

Ballast lines can be measured by changing the coldest spot temperature. The parts of the ballast lines below the operating point are measured during the

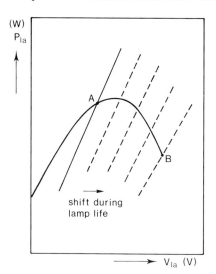

Figure 6.21 Shift of the operating point of an HPS lamp during lamp life. The full lines show a typical ballast line and a lamp line for a new lamp, called an initial lamp line. Because of voltage rise during lamp life, the lamp line shifts to higher voltage values as shown by the dashed lines. The operating point of the HPS lamp – for a new lamp point A – can always be considered as the intersection of a ballast line and a lamp line. Point B is the extinction point.

run-up phase of the lamp, while the parts above the operating point are measured with the lamp wrapped in aluminium foil. When the lamp voltage reaches the extinction point, the necessary reignition voltage exceeds the instantaneously available mains voltage; operation at higher coldest spot temperatures is not then possible.

The lamp lines for a given ballast can be measured by varying the supply voltage so as to change the lamp power. Because of the lamp voltage rise during lamp life, the lamp lines shift to higher lamp voltages for older lamps, as is illustrated in figure 6.21.

In the following sections it will be explained how the demands placed on the stabilisation circuits are met with various types of circuits.

6.2.2 Capacitive and Inductive Stabilisation

Calculated Ballast Lines
Ballast lines and lamp lines can be helpful in characterising the main properties of ballasts. This section will be limited to a consideration of the main properties of two basic types of ballasts – the inductive and the capacitive ballasts shown in the circuits of figure 6.22.

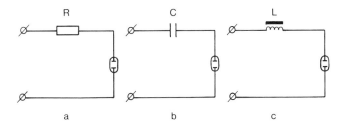

Figure 6.22 The three basic ballasts: (a) resistive (b) capacitive and (c) inductive.

In the ideal circuit of pure capacitive or inductive stabilisation with a sinusoidal 50 Hz power supply, the lamp power is determined by the product of the 50 Hz component of the lamp voltage ($V_{\mathrm{la,50}}$), the 50 Hz component of the lamp current ($I_{\mathrm{la,50}}$), and the cosine of the phase angle θ between these two components

$$P_{\mathrm{la}} = P_{\mathrm{la,50}} = V_{\mathrm{la,50}}\, I_{\mathrm{la,50}} \cos\theta \qquad (6.7)$$

Higher frequency harmonics occur in the lamp voltage and current in the case of the capacitive (lead) and inductive (lag) types of stabilisation. These higher frequency harmonics generated by the discharge lamp do not dissipate power in a purely capacitive or inductive circuit and so do not contribute to the lamp power. In practice, of course, ballasts are not free of losses, and the resultant lamp power will be smaller than that determined by the 50 Hz component ($P_{1a} < P_{1a,50}$). The lamp transforms 50 Hz power from the mains into higher-harmonic losses in the ballast.

The lamp power can easily be calculated with the aid of vector diagrams (figure 6.23) for the 50 Hz components of voltage and current in the ideal, simple circuits of figure 6.22. From these diagrams it follows that for the 50 Hz component $I_{1a,50}$ of the lamp current

$$I_{1a,50} = \frac{[V_m^2 + V_{1a,50}^2 - 2\,V_m\,V_{1a,50}\cos(\varphi - \theta)]^{1/2}}{Z_{50}} \tag{6.8}$$

with

$$\varphi = \text{arc cos } \frac{V_{1a,50}}{V_m}\cos\theta$$

where φ = phase angle between the mains voltage and the 50 Hz compo-
              nent of the lamp current
       θ = phase angle between the 50 Hz-components of the lamp volt-
              age and lamp current
       V_m = mains voltage
       $V_{1a,50}$ = the 50 Hz component in the lamp voltage
       Z_{50} = ballast impedance at 50 Hz

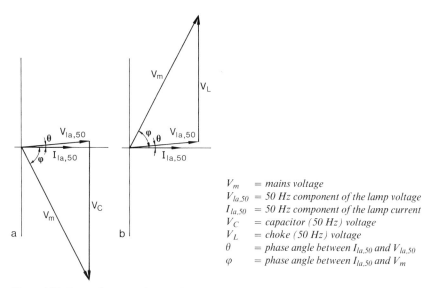

V_m = mains voltage
$V_{la,50}$ = 50 Hz component of the lamp voltage
$I_{la,50}$ = 50 Hz component of the lamp current
V_C = capacitor (50 Hz) voltage
V_L = choke (50 Hz) voltage
θ = phase angle between $I_{la,50}$ and $V_{la,50}$
φ = phase angle between $I_{la,50}$ and V_m

Figure 6.23 Vector diagrams showing the 50 Hz components in the idealised (a) capacitive and (b) inductive circuits of figure 6.22. The lamp current is assumed to be lagging with respect to lamp voltage.

Where the lamp current lags the mains voltage, φ is positive, and where it leads, φ is negative. The lamp power depends on the mains voltage and the lamp voltage as follows

$$P_{la,50} = \frac{V_m^2 f \cos [1 + f^2 - 2 f \cos (\varphi - \theta)]^{1/2}}{Z_{50}} \tag{6.9}$$

where $\quad f = \dfrac{V_{la,50}}{V_m}$

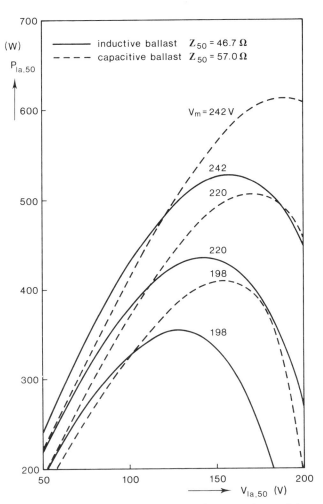

Figure 6.24 Calculated ballast lines for ideal inductive and capacitive lamp stabilisation. The lamp power ($P_{la,50}$), calculated according to Eq. (6.9) is given as a function of the 50 Hz component of the lamp voltage $V_{la,50}$ for an inductive ballast (Z_{50} = 46.7 Ω) and a capacitive ballast (Z_{50} = 57.0 Ω) for three mains voltage values.

The lamp current lags the lamp voltage slightly, so θ is positive. The phase angle θ between the 50 Hz components of the lamp voltage and the lamp current cannot be accurately determined. For a 400 W HPS lamp, the measured value for θ is about $10°$, and θ is not seriously affected by variations of mains voltage and lamp voltage. Eq. (6.9) has been used to calculate the influence of the 50 Hz component of the lamp voltage on the lamp power (figure 6.24). The following conclusions can be drawn from these idealised ballast lines

– With respect to lamp voltage variations, the inductive ballast shows a better power regulation than does the capacitive ballast, especially for a lamp voltage of about half the mains voltage.
– At a constant lamp voltage (about half the mains voltage) the capacitive ballast regulates the lamp power against mains voltage fluctuations somewhat better than does the inductive type.
– The higher the lamp voltage, the larger the lamp power changes due to mains voltage fluctuations. These changes in lamp power can thus only be reduced by making the lamp voltage small with respect to the mains voltage; in other words, where the voltage across the ballast is such as to give a relatively large volt-ampere value for the ballast.

It will be clear that the aims of reducing the influence of mains voltage fluctuations on the one hand and ballast losses on the other are mutually conflicting. It has been found in practice that a lamp voltage of about half the mains voltage provides a good compromise solution to this problem.

Availability of Reignition Voltage
With either capacitive or inductive stabilisation there is a phase-shift between the mains voltage and the lamp current, as shown by the oscillograms in figure 6.25. Therefore, in contrast with resistive stabilisation (see Sec. 6.2.3), there will be a voltage available for reignition at current zero. However, as the mains voltage falls, the voltage available for reignition decreases. If the lamp voltage increases during the life of the lamp, the voltage necessary for reignition will increase as well. The problem of lamp extinction therefore becomes more serious if, at the end of lamp life, sudden dips in the mains voltage should occur. If this is the case, then because of the negative voltage-current characteristic, the lamp voltage will increase before the lamp stabilises at a lower lamp voltage. Figure 6.26 shows the ballast lines of an inductive ballast stabilising a 400 W HPS lamp for three values of the mains voltage, and the initial lamp line of this lamp. The dotted line marked 0% is the extinction line; it gives the lamp voltages above which operation of the lamp is not possible. The lamp line shows that for a lower mains voltage the lamp voltage finally stabilises on a lower value. However, when a sudden mains voltage dip occurs, the lamp current will be abruptly reduced – and because

198

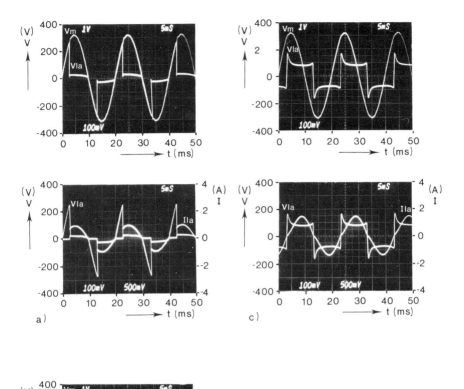

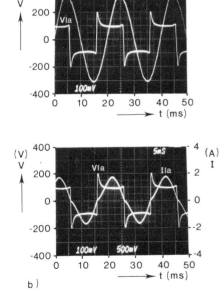

Figure 6.25 Oscillograms of lamp current I_{la}, lamp voltage V_{la} and mains voltage V_m as functions of time for a 70 W HPS lamp stabilised on (a) a resistive (b) a capacitive and (c) an inductive ballast.

199

of the negative voltage-current characteristic of the lamp – the lamp voltage will at first rise. Extinction will therefore occur at a lower lamp voltage with mains voltage dips than with a steady mains voltage. The dotted lines marked –5% and –10% in figure 6.26 are the extinction lines for 5% and 10% voltage dips respectively. During the life of the lamp, the lamp line shifts to higher lamp voltages. Extinction due to mains voltage fluctuations occurs if the lamp line intersects the corresponding extinction line.

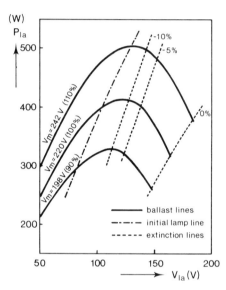

Figure 6.26 Ballast and extinction lines for a 400 W HPS lamp stabilised on a choke ballast. The initial lamp line and extinction lines at mains voltage dips of 0, –5 and –10 per cent are also given. If, during the life of the lamp, the lamp line shifts to higher lamp voltage values and intersects the relevant extinction line, then extinction will occur during corresponding fluctuations of the mains voltage.

The effect of the buffer gas xenon on the electrical behaviour near current zero was illustrated in Sec. 5.4. An increase in pressure of the xenon starting gas lowers the reignition voltage (van Vliet and Nederhand, 1977; Iwai *et al.*, 1977). An HPS lamp with a high xenon pressure can therefore withstand more violent dips in the mains voltage than can an HPS lamp with xenon at low pressure.

Additional Remarks on Capacitive and Inductive Stabilisation
The purely capacitive stabilising element produces an unacceptable distortion of the mains current, the capacitor presenting a virtual short-circuit to the higher frequency harmonics in the current. One way to avoid such distortion would be to employ a capacitive ballast circuit comprising a combination of capacitor and choke (combined capacitive circuit), the latter having about half the reactance of the former at the mains frequency. However, this circuit can best be used in conjunction with discharge lamps whose lamp voltage

is nearly independent of the lamp power, e.g. lamps such as the HPMV lamp. This is because the ability of this circuit to compensate for changes in the lamp voltage is less good. Although it can ensure a high reignition voltage and also good regulation of the lamp power during mains voltage fluctuations, it is not advisable to use the capacitive-type ballast in conjunction with HPS lamps. The reason for this becomes clear if one considers the nature of the lamp lines for typical HPS lamps and the ballast lines belonging to the combined capacitive circuit. In figure 6.27 ballast lines are given for a 350 W HPS lamp operated on a combined capacitive and inductive ballast at mains voltages of 220 V and 242 V. Also given are two lamp lines, the slight shift between which is thought to be due to either initial production tolerances or changes occurring during the life of the lamp. The capacitive stabilising circuit acts as a constant-current source for the lamp. Since the lamp and ballast lines are nearly parallel, it is not practicable to stabilise the lamp power with this circuit, for soon after switching on the mains the lamp power, and also the lamp voltage, will increase until the lamp extinguishes. Capacitive ballasts are used, in countries where the mains voltage is 120 V, in combination with a special transformer (Unglert and Kane, 1980; McVey and Paugh, 1982) to step up the 120 V mains to a higher voltage.

As can be seen from the shape of the ballast lines in figure 6.27, the inductive ballast promises better stabilisation than does the capacitive one with respect to lamp voltage changes. For the supply frequency of 50 Hz a relatively heavy choke or transformer has to be used to keep the iron and copper losses within acceptable limits. In table 6.1 the losses of commercially available ballasts for various types of HPS lamps are given. They amount to 4 to 8 per cent of the volt-ampere product of the ballast.

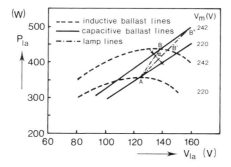

Figure 6.27 Ballast lines for a 350 W HPS lamp operated on a combined capacitive and an inductive ballast at mains voltages of 220 V and 242 V, together with two lamp lines shifted by an angle α with respect to one another. With the inductive ballast, the increase in lamp power at a 10 per cent higher V_m is nearly independent of lamp-line shift ($B \to B'$). For a combined capacitive ballast, however, the increase in lamp power for the same mains voltage variations is strongly dependent on lamp-line shift ($B \to B''$). (de Neve, 1976)

201

An advantage of inductive stabilisation is that a starting circuit can easily be incorporated in the choke ballast to produce the high-amplitude ignition pulses needed for HPS lamps. The disadvantage of the inductive ballast with respect to large mains voltage fluctuations can be avoided by using another type of lag ballast, the so-called regulated, or constant-wattage type of ballast. In this device, a magnetic voltage-regulator transformer is added to the choke. Such a transformer gives a stabilised voltage on its secondary windings.

Table 6.1 Losses in commercial ballasts according to the type of lamp employed (220V/50Hz mains voltage)

HPS lamp type	Ballast losses (W)
50 W	8
70 W	11
100 W	11
150 W	20
250 W	30
400 W	35
1000 W	60

For a practical V_{la}/V_m ratio of 0.5 with inductive stabilisation, resulting in a phase-shift of about 60°, the current waveform will be nearly sinusoidal. However, with a phase-shift of 60°, the power factor of the circuit will be below the 0.85 desired by the electricity supply authority. The power factor may be increased by connecting a capacitor across the mains to provide compensation for the lagging component of the 50 Hz lamp current.

6.2.3 Resistive Stabilisation

Resistors are not commonly used for stabilisation, mainly because of the high power dissipation in such a ballast, which reduces the overall luminous efficacy of the discharge lamp and ballast combination. The only advantage of using a resistor as a ballast is that it will be light and cheap, especially if it is in the form of an incandescent filament incorporated in the lamp – as in the so-called blended-light lamp. An example of such a blended-light lamp is the lamp in which the light of a high-pressure mercury discharge lamp is coupled with that of a filament (Elenbaas, 1965). Since, in the case of resistive stabilisation, current and mains voltage will be in phase, reignition will be hampered. This can be seen from the oscillograms of current, and lamp and mains voltage shown in figure 6.25a. Reignition does not take place before the mains voltage has increased sufficiently for this to happen. At the end of every cycle, as the mains voltage begins to decrease to a value below the lamp voltage, the discharge extinguishes. The lamp current and voltage waveforms will therefore be heavily distorted, which results in a low power factor. Also, the long dark periods can produce undesirable flickering of the light output.

But the main disadvantage of the resistive ballast remains its relatively high power dissipation. The luminous efficacy η of the blended-light lamp is given by

$$\eta = \eta_d \frac{P_d}{P_m} + \eta_f (1 - \frac{P_d}{P_m}) \tag{6.10}$$

where η_d = luminous efficacy of the discharge (including electrode power)
          η_f = luminous efficacy of the filament
          P_d = power dissipated in the discharge (including electrode power)
          P_m = power supplied to the blended lamp by the mains

Because the luminous efficacy of the filament is very low ($\eta_f \sim 7$ to 15 lm W^{-1}) with respect to that of the discharge, the luminous efficacy of the blended-light lamp is roughly proportional to the ratio P_d/P_m, which is approximately equal to the ratio of the discharge voltage to the mains voltage. Therefore, in the case of the blended-light lamp, the maximum possible luminous efficacy is achieved by ensuring that the discharge voltage is made as high as possible with respect to the mains voltage. This ratio is limited, however, by the electrical behaviour near current zero and by the need to limit the fluctuations in light output for a given mains-voltage fluctuation.

Raising the discharge voltage causes the discharge to extinguish at the moment the reignition peak 'exceeds' the instantaneous value of the mains voltage. Therefore, in order to obtain an efficient blended-light lamp, the difference between reignition voltage and discharge voltage after reignition must be minimised. This can be achieved by increasing the relaxation time for plasma cooling. The higher the pressure and the lower the thermal conductivity of the buffer gas and the larger the diameter of the discharge tube, the greater is the relaxation time (Eq. 4.35). In this way higher values for P_d/P_m can be obtained. Figure 6.28 shows this fraction as a function of the ratio of the mercury buffer gas pressure to the sodium vapour pressure at a constant $\Delta\lambda$–value for two tube diameters. In order to ensure that the $\Delta\lambda$–value was approximately constant on each curve and the same for both diameters, the composition of the sodium amalgam in the test lamps and the coldest spot temperature were adjusted accordingly. Still higher values for P_d/P_m can be obtained at high xenon pressures and low sodium vapour pressures (Sec. 5.4).

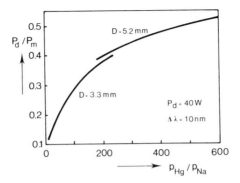

Figure 6.28 The ratio of discharge power P_d (including electrode losses) to mains power P_m as a function of the ratio of the mercury to the sodium vapour pressures, p_{Hg}/p_{Na}, for discharge tube diameters of 3.3 mm and 5.2 mm. The wavelength separation between the maxima of the D-lines is held constant at a value of 10 nm. The discharge power is held constant at 40 W.

However, for a discharge voltage of more than half the supply voltage, the variation in lamp current with increase or decrease in mains voltage will be unacceptably large. In practice P_d/P_m has to be kept to less than 0.5. So, were an incandescent filament to be used for stabilisation, the luminous efficacy of the blended-light lamp would be about half of the HPS discharge lamp itself.

At the time of writing, no HPS blended-light lamps are in production. This is in spite of the improvement in luminous efficacy that such a lamp would exhibit as compared with the present blended-light lamps. The reason is that the HPS version of the blended-light lamp would be more expensive than the HPMV version. Another possible way of replacing incandescent lamps or blended-light HPMV lamps by HPS lamps is to use an adapter with a built-in ballast of the inductive type together with a starter (Wyner *et al.*, 1982).

6.2.4 Quadrilateral Diagram

Because an excess of amalgam is present in the HPS lamp, a change in amalgam temperature or composition is coupled with a change in the sodium and mercury pressures (brought about by either evaporation or condensation of both elements), and so too with a change in the lamp voltage. Such changes may be caused by (Jacobs *et al.*, 1978; Inouye *et al.*, 1979)

a) a clean-up of sodium during lamp life (which is the reason for the excess of amalgam in HPS lamps); the amalgam becomes richer in mercury, so producing a higher mercury pressure and a lower sodium pressure at the same amalgam temperature;

b) a change in the (amalgam) temperature at the coldest spot during lamp life, for which there are several contributory factors:
   – blackening of the end of the discharge tube by deposits of sputtered or evaporated electrode material;
   – a changed power flow through the electrode.
Also, the initial lamp voltage shows a spread because of
a) production tolerances in the discharge lamp;
b) production tolerances in the ballast;
c) the influence of the luminaire in which the lamp is operated, each luminaire having its own specific optical and heat insulating properties;
d) the influence of mains voltage fluctuations.

Ballast and lamp manufacturers therefore impose certain limitations on their products so as to ensure that they operate within specification. The lamp operating limits can be conveniently specified by means of a quadrilateral or trapezoid diagram (Davies, 1980; Unglert and Kane, 1980; Unglert, 1982),

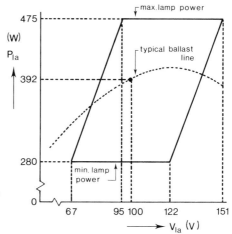

Figure 6.29 Lamp operating limits for a 400 W HPS lamp given in the quadrilateral diagram. (IEC, 1979)

as given in figure 6.29. The upper boundary defines the maximum permissible power in the lamp, for which its lifetime is still reasonable. The lower boundary, marking the minimum permissible power in the lamp, is to ensure an acceptable luminous flux after warm-up and a satisfactory warm-up time. The position of the left-hand boundary, which is marked by a lamp line, is not so critical. Working within this boundary can provide an indirect protection against excessive lamp current. The right-hand boundary is also marked by a lamp line. The extinction line for steady mains voltage should lie well outside this boundary.

205

In selecting the lamp circuit and fixing the design of the ballast, the ballast designer will take good care that fluctuations in the supply voltage do not cause the ballast line to intersect the lower or upper boundaries of the quadrilateral diagram. The lamp designer has to keep the initial value for the lamp voltage to the right of the left boundary, taking care that the initial variations in lamp voltage are as small as possible. He must also ensure that the lamp voltage does not cross the right-hand boundary during the life of the lamp.

6.2.5 Solid-State Ballast

The most important reason for the interest in solid-state ballasts is that they are smaller and lighter than conventional ballasts. In addition, those employed in conjunction with the low-wattage HPS lamps are more efficient, their power consumption being lower than that of conventional ballasts (Verderber and Morse, 1982). Also, the solid-state ballast opens the way to high-frequency or pulse-current operation (Secs 7.1 and 7.2). In the case of pulse-current operation, a higher correlated colour temperature can be obtained than with conventional ballasts, while at higher frequencies, reignition problems can be eliminated. Solid-state ballasts are also better able to regulate lamp power, lamp voltage and light output.

The schematic diagram of figure 6.30 shows the main circuit components of a possible solid-state ballast, along with their functions. The low-pass filter corrects for mains-current distortion, suppresses radio-frequency interference (r.f.i) and removes spikes on the mains to protect the electronic circuitry.

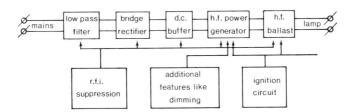

Figure 6.30 Block diagram of a possible solid-state ballast. Additional circuits may be added to suppress radio interference and mains-current distortion or to make ignition or dimming possible. (Rozenboom, 1983)

The bridge rectifier, in combination with the d.c. buffer, converts the mains voltage into a d.c. voltage. The high-frequency power generator converts the d.c. voltage into a high-frequency voltage on which the lamp, in series with a h.f. choke, is operated.

The solid-state ballast containing only switching elements cannot perform all the functions of the copper-iron ballast. In the first place, discharge lamps cannot be operated properly without energy-storage elements to aid starting and stabilisation. Secondly, the current periodically interrupted by the semi-conductor has to be smoothed by storage elements to avoid radio interference. The solid-state ballast therefore generally contains both active and passive elements, which determine the size and weight of the unit. The solid-state ballast can also be used to electronically dim the lamp in order to reduce energy consumption (van Os, 1982).

Finally, where solid-state ballasts are involved, the ballast designer and lamp designer have to work in close collaboration with one another because the interaction between ballast and lamp is generally stronger than with the conventional ballast. They have to carefully choose lamp parameters such as lamp current and voltage, along with frequency and waveform, to get the best overall results for the system, without creating radio interference or acoustic resonance problems (Sec. 7.3).

Chapter 7

High-Frequency and Pulse Operation

New electronic components (e.g. high-power transistors) have been developed that make it possible to operate HPS lamps on currents widely different in frequency and waveform to those present in HPS lamps operated on a conventional, mains-frequency ballast. The electronic ballast makes it possible to reduce the volume and weight of the stabilising circuit and to improve the system efficiency. Furthermore, the regulation of the lamp power at a constant value in the event of lamp voltage rise and the regulation of the light level can also be taken care of by the solid-state ballast (Verderber and Morse, 1982). Other properties of the lamp can also be positively or negatively influenced, as will be discussed in this chapter.

The influence of the frequency and waveform of the supply current on the properties of the HPS discharge is closely connected with the relaxation time for plasma cooling; viz. if the period of the current waveform exceeds this relaxation time, the plasma temperature will be modulated. In the opposite case, where this period is much less than the relaxation time, the plasma temperature will be nearly constant during the period, as in the d.c. situation. The modulation depth of the plasma temperature will influence the various discharge properties in different ways, as will be discussed for high-frequency sinusoidal and pulse operation in Secs 7.1 and 7.2 respectively. A disadvantage of high-frequency or pulse operation may be the occurrence of discharge distortions by acoustic waves. This will be discussed in Sec. 7.3.

7.1 High-Frequency Sinusoidal Operation

Figures 7.1 to 7.7 give the measured and calculated operating characteristics of a 70 W HPS lamp over the frequency range 50 Hz–20 kHz. The measured data (Dorleijn and van der Heijden, 1980a) were recorded at frequencies where no visible disturbance due to acoustic resonances was observed (Sec. 7.3). For the calculation of the corresponding data, use is made of the non-steady-state LTE-model described in Chapter 4. From these figures we can conclude that as the frequency of the sinusoidal discharge current increases, then

208

a) The reignition and extinction peaks disappear (figure 7.1).
b) The lamp power factor approaches unity (figure 7.2).
c) The amplitude modulation of the light output decreases and the phase shift between the time of minimum light output and current zero increases (figures 7.3 to 7.5).
d) The spectral power distribution is not significantly altered (figure 7.6);

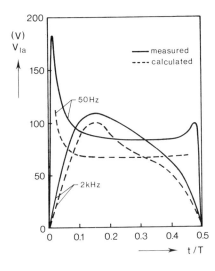

Figure 7.1 Measured and calculated lamp voltage V_{la} for a 70 W HPS lamp as functions of the reduced time t/T during one half period of the 50 Hz and 2 kHz supply current. The difference between calculated and measured (lamp) voltages is partly due to the fact that the electrode fall was neglected in the calculation. The reignition and extinction peaks at the beginning and the end of the half cycle disappear for high-frequency operation.

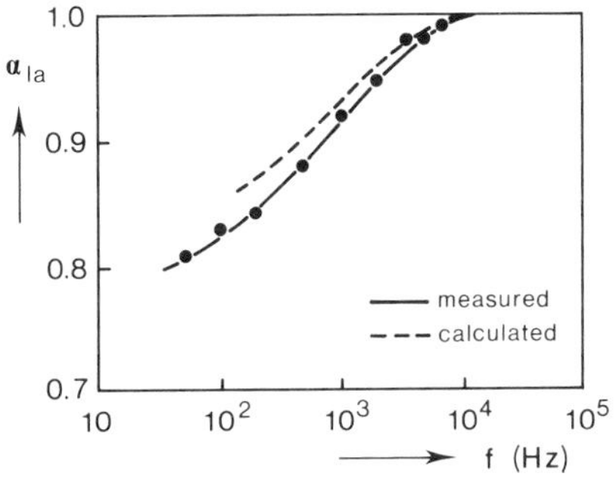

Figure 7.2 Power factor of a 70 W HPS lamp over a range of supply current frequencies.

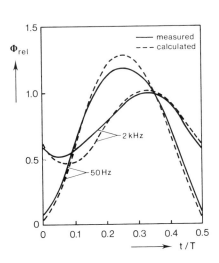

Figure 7.3 Relative luminous flux (in arbitrary units) of a 70 W HPS lamp as a function of the reduced time t/T during one half-period of the 50Hz and 2 kHz supply current.

209

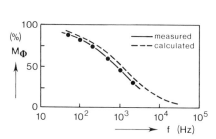

Figure 7.4 Modulation depth of the luminous flux from a 70 W HPS lamp over a range of supply current frequencies.

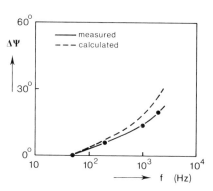

Figure 7.5 Phase shift between minimum light output and current zero for a 70 W HPS lamp over a range of supply current frequencies.

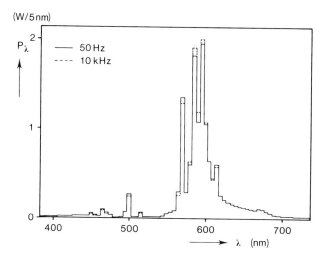

Figure 7.6 Measured spectral radiant power for a 70 W HPS lamp operated at 50 Hz and 10 kHz.

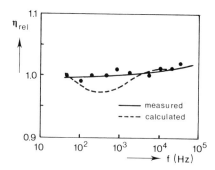

Figure 7.7 Relative luminous efficacy of a 70 W HPS lamp over a range of supply current frequencies.

210

it is for this reason that the relative change in the radiation efficiency is about the same as the relative change in the luminous efficacy.

e) Up to 20 kHz the luminous efficacy of the lamp remains almost constant[*] (figure 7.7).

These characteristics of HPS lamps operated at high frequencies (sinusoidal waveform operation) can be understood from the time-dependent behaviour of the plasma temperature during one half-period of the supply voltage.

7.1.1 Plasma Temperature Modulation

The modulation of the plasma temperature in the hot inner channel of the discharge can be deduced from the modulation depth of optically thin spectral lines. The effect of the plasma temperature modulation is found by considering an LTE plasma of spatially constant temperature T. The radiant intensity I_L of an optically thin line L is related to the temperature T and the energy E_u of the relevant upper level by

$$I_L \sim \frac{1}{T} \exp(-\frac{E_u}{kT})$$

(7.1)

The modulation depth M_I of the intensity of the spectral line can be written as

$$M_I = \frac{I_{Lmax} - I_{Lmin}}{I_{Lmax} + I_{Lmin}} = \tanh\left[\frac{\Delta T}{T}(\frac{E_u}{kT} - 1)\right]$$

(7.2)

provided that $\dfrac{\Delta T}{T} \ll 1$

where $T = \dfrac{T_{max} + T_{min}}{2}$

and $\Delta T = \dfrac{T_{max} - T_{min}}{2}$

In the actual discharge, the radiation of an optically thin line comes almost entirely from the hot central core. Therefore, the modulation depth of the intensity can, knowing the average plasma temperature near the discharge axis, be used to derive the temperature modulation from Eq. (7.2).

The modulation depth of the plasma temperature as measured by Dorleijn and van der Heijden (1980a) is compared in figure 7.8 with the modulation

---

[*] This is in apparent contradiction with the experimental results of Campbell (1969) and Verderber and Morse (1982), who reported 8 per cent and 4 per cent increases in luminous efficacy, respectively, for a 400 W HPS lamp at 20 kHz with respect to that for 60 Hz operation.

depth calculated from the time-dependent, power-balance equation for a LTE plasma as described in Chapter 4 – see Eq. (4.7). Good agreement is found between the measured and calculated modulation depths up to a frequency of about 10 kHz. The deviation above this frequency is caused by the fact that the time-dependent LTE model considers only the relaxation of kinetic and potential energy of the gas atoms due to conduction and radiation losses. The model disregards the relaxation effects of energy transfer between the electrons on the one side and the atoms on the other. The frequency dependence of the optogalvanic effect as measured in high-pressure discharges by van den Hoek and Visser (1980) is also explained, for example, with such energy relaxation effects.

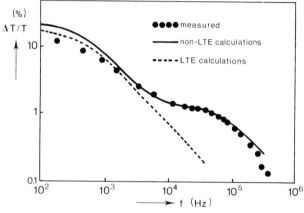

Figure 7.8 Modulation depth of the plasma temperature in a 70 W HPS lamp over a range of supply current frequencies. Two calculated curves are presented, one of which is based on an LTE model (Chapter 4), the other on a non-LTE model (Sec. 7.1.2). (Dorleijn and van der Heijden, 1980a)

From the data presented in figure 7.8 we see that the modulation depth of the plasma temperature is reduced from 20 per cent at 50 Hz to 1 per cent at 10 kHz. The modulation in light output (figure 7.4) and the reignition and extinction peaks will disappear as the frequency of the supply current is increased, resulting in a lamp power factor of unity at high frequency (figures 7.1 and 7.2). Thus, at increasing frequencies the operating characteristics of the HPS lamp would appear to approach those found under d.c. operation. Measurements on d.c. operated HPS lamps are hampered by cataphoretic effects. Such operation results in demixing of the gas-filling, as is illustrated in colour plate 6b. The sodium is transported toward the cathode side of the tube. By changing the polarity of the supply – for example every 10 ms – an axially homogeneous discharge is obtained.

7.1.2 Non-Steady-State Non-LTE Model

Dorleijn and van der Heyden (1980b) explained the frequency dependence of the plasma temperature modulation depths and phase shifts found experimentally above 10 kHz with the help of separate, non-steady-state power balance equations for the atoms and the electrons. In formulating their strongly simplified model, Dorleijn and van der Heyden considered a cylindrical discharge channel in which all radial variations of temperatures, particle densities, etc. were neglected and replaced by radially constant values. The chief physical features of their discharge model were as follows

a) The supply of electric power to the discharge results in a heating of the electron-gas, which transfers power to the heavy particles in elastic and inelastic collisions. The bulk of the power is transferred as heat by elastic collisions. The portion transmitted in the form of internal energy of atoms by inelastic collisions, is neglected. The neutral particles lose heat by conduction and radiation.

b) The energy-level diagram of the atom is simplified to a two-level diagram. All the radiated power is assumed to be emitted from one imaginary upper level. The radiant power is assumed to be proportional to the number of excited atoms present in the discharge.

Considering only small temperature fluctuations around an average value, two coupled differential equations with four adjustable parameters were derived, describing the separate power balances for the electrons and atoms. By the appropriate choice of the four parameter values, called τ_e, τ_g, c_1 and c_2, a calculated curve was obtained that passed through the measured values. Figures 7.8 and 7.9 give the best-fit curves obtained by the method of least-squares for the modulation depth and the phase shift using the values of the two parameters τ_e, τ_g listed in table 7.1, which parameters may be inter-

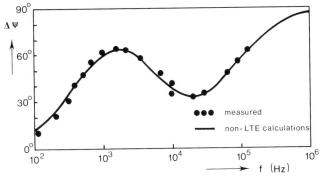

Figure 7.9 *Measured phase-shift between minimum plasma temperature and current-zero for a 70 W HPS lamp over a range of supply current frequencies. Also given is the calculated best-fit curve for the phase-shift, based on the non-LTE channel model as described in Sec. 7.1.2.*

preted as the characteristic times for energy relaxation of the gas atoms and the electrons respectively. The time-dependent behaviour of the gas discharge for low frequencies is mainly governed by the energy relaxation time τ_g. This relaxation time increases with increasing pressure as shown in table 7.1.

Table 7.1 Values for relaxation times τ_g and τ_e (see Sec. 7.1.2) used to obtain the best-fit curves for the modulation depth and phase shift as measured for various 70 W HPS lamps (D = 3.3 mm)

	lamp 1	lamp 2	lamp 3
p_{Na} (kPa)	39	15	12
p_{Hg} (kPa)	0	146	118
p_{Xe} (kPa)	3	35	472
τ_g (μs)	18	60	175
τ_e (μs)	1.5	1.3	1.3

Thus, although the discharge model as used by Dorleijn and van der Heijden is perhaps oversimplified, it does provide an explanation of the results for modulation depth and phase shift at higher frequencies.

7.2 Pulse-Current Operation

The general object of pulse-current operation of HPS lamps is to improve the colour quality of the light emitted (Osteen, 1979). It offers the possibility of increasing the peak current – and thus the instantaneous input power – during a small fraction of the current period. The result is an instantaneously high plasma temperature and a corresponding change in the spectrum.

The plasma temperature is increased by two effects: the 'dynamic' effect and the 'static' effect. If the time between two successive pulses is relatively large, the plasma will cool sufficiently between the pulses to produce 'reignition' through a narrow channel at every pulse. This will lead to high axis temperatures at the moment of reignition, as has already been shown in Sec. 4.2.3 – the so-called dynamic effect.

For a relatively large pulse width, the effect on the time-averaged properties is slight. This is because of the rapid broadening of the discharge channel, which results in an axis temperature lower than that at the moment of reignition. The dynamic effect is completely absent if the plasma cannot cool sufficiently between two pulses to allow reignition through a narrow channel. In this case, only the 'static' effect is present; as a result of the increased power input, the axis temperature will be raised towards a steady-state situation (Sec. 4.2.1). We shall start by considering the static effect.

214

7.2.1 Static Effect

The static effect was investigated by feeding a 400 W HPS lamp with positive and negative high-amplitude current pulses, both with a repetition frequency of 50 Hz, as is shown schematically in figure 7.10. During the rest of the 50 Hz period a small, so-called keep-alive current was flowing, so the reignition peak voltage was less than the maximum value delivered by the electrical circuit. Superimposed on this keep-alive current was a current pulse of such amplitude and width that the time-averaged power was held constant at 400 W. The influence of the dynamic effect was reduced by avoiding strong reignition phenomena with the keep-alive current and by making the pulse width large with respect to the relaxation time of the gas-kinetic energy reservoir.

The spectral power distributions for pulse-current amplitudes of 5 A and 20 A are shown in figure 7.11. As already stated in Sec. 3.1, at the higher

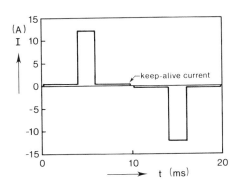

Figure 7.10 Current waveform composed of a current pulse superimposed on a keep-alive current. The amplitude and width of the current pulse can be varied so that the time-averaged power is held at a constant value.

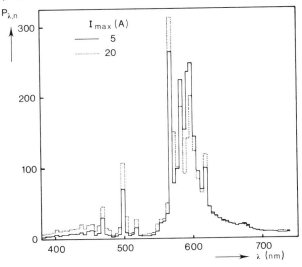

Figure 7.11 Time-averaged measured spectral power distribution (normalised per unit of luminous flux) for a 400 W HPS lamp at pulse-current amplitudes of 5 and 20 A.

215

axis temperature the increased recombination radiation and strongly Stark-broadened line radiation from the high-lying d-levels give increased radiation in the blue region of the spectrum with respect to the D-line radiation. This phenomenon is responsible for the higher colour temperature and whiter colour appearance of the lamp as compared to that obtained with d.c. or sinusoidal operation. The influence of the increased radiation of mercury lines at the high current value is of minor importance.

In figure 7.12, the time-averaged values for the luminous efficacy, the correlated colour temperature and the general colour rendering index are given as functions of the maximum current amplitude during the 50 Hz period.

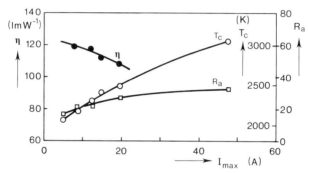

Figure 7.12 Time-averaged measurements of luminous efficacy η, general colour rendering index R_a, and correlated colour temperature T_c for a pulse-operated (50 Hz) 400 W HPS lamp as functions of the maximum current I_{max} within the pulse.

From these data it can be concluded that the colour temperature can be increased to 3000 K at the expense of a decrease in luminous efficacy of more than 20 per cent, which is in agreement with the results of Osteen (1979). The general colour rendering index increases only slightly with this increase in current-pulse amplitude.

By solving the steady-state power balance equation (Chapter 4), the plasma temperature distribution can be calculated for several current values. The calculated axis temperatures are given in figure 7.13 as a function of the current through a 400 W HPS lamp. These calculations are strictly speaking only valid for a d.c. situation. But for pulse-widths larger than several hundred microseconds – thus larger than the energy relaxation time of the gas atoms – these steady-state temperature distributions will approximate the real temperature distributions reasonably well. Figure 7.13 also gives the correlated colour temperatures obtained from the calculated spectra (Secs 3.1 and 5.2) corresponding to these calculated temperature distributions. The calculated increase of the colour temperature agrees reasonably well with the measured increase for pulse-operated HPS lamps (figure 7.12).

216

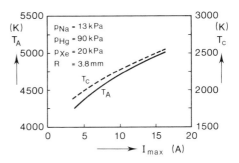

Figure 7.13 Calculated axis temperature T_A and colour temperature T_c as functions of the pulse current amplitude I_{max} of a pulse-current-operated 400 W HPS lamp. The pulse duration is assumed to be large with respect to the time for energy relaxation of the gas atoms.

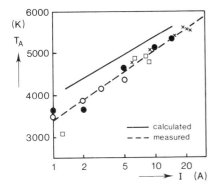

Figure 7.14 Measured axis temperature at peak light output of a pulse-current-operated 150 W HPS lamp as a function of the pulse-current amplitude (Johnson and Rautenberg, 1979)

- • 60 Hz sine wave
- × 60 Hz pulsed
- □ 1 kHz square wave, 30% duty cycle
- ○ 10 kHz gated at 600 pps, 20% duty cycle

The calculated axis temperature (based on the steady-state LTE model described in Chapter 4) as a function of discharge current is also given (solid line).

7.2.2 Dynamic Effect

If HPS lamps are operated with current-pulse durations smaller than the gas relaxation time τ_g and repetition times larger than τ_g, the maximum plasma temperature of the hot core of the discharge will be greatly increased. As was shown in Chapter 4, if the keep-alive current is omitted, at reignition the current is forced to flow through a narrow conductive core of the discharge (figure 4.19). It is very difficult to make use of the dynamic effect in operating HPS lamps on pulse-shaped lamp currents because of the high reignition peaks connected with this effect. Therefore only combined static and dynamic effects are reported in literature.

Pulse-current operation at frequencies above 50 Hz has been studied by Johnson and Rautenberg (1979), by Chalek and Kinsinger (1981) and by Dakin and Rautenberg (1984). Johnson and Rautenberg measured the maximum axis temperature for a 150 W HPS lamp under pulsed conditions as a function of the instantaneous current at maximum light output (figure 7.14). Chalek and Kinsinger ascribe the high plasma temperatures for the high peak-current values to a narrow channelling of the arc current through the centre of the arc. From a comparison – at equal current values – of these measured axis temperatures as a function of the instantaneous current at maximum light output, and our calculated values as a function of the discharge current under steady-state conditions (figure 7.14), it is concluded that the increase in axis temperature is due mainly to the static effect. This means that the dynamic effect is less important than the static effect under these conditions.

7.3 Acoustic Resonances

The operation of high-pressure discharge lamps on high-frequency current waveforms is hampered by the occurrence of standing pressure waves (acoustic resonances) (Scholz, 1970), which may lead to discharge-path distortions. These acoustic resonances in HPS lamps were first reported by Witting

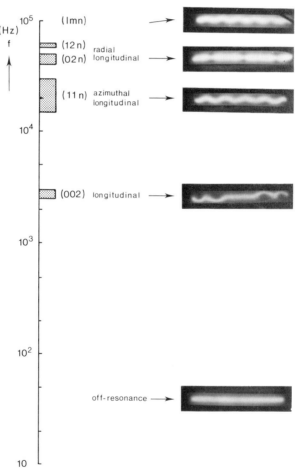

Figure 7.15 Photographs of the discharge paths in a 250 W HPS lamp at various frequencies of the current supply. The letters (l, m, n) indicate the mode of resonance occurring in the hatched frequency region according to Eq. (7.8).

218

(1978). The results presented in this section stem mostly from the work of
Dorleijn and van der Heijden (1980b).

The various visual appearances of the discharge path in a 250 W HPS lamp
when the sinusoidal lamp current frequency is increased from 50 Hz to 100
kHz, are given in figure 7.15. The photographs in this figure show some distorted but stable discharge paths observed in various frequency bands (the
hatched areas). Outside these bands the discharge has the same appearance
as when operated at 50 Hz. These distortions are caused by the occurrence
of acoustic resonances, as will be shown in this section.

These observations are typical for all types of HPS lamps. The most remark-
able difference between various lamp types is that the bands may be shifted
to other frequencies due to the different discharge-tube dimensions. Also,
a different number of bands can occur. This can be explained by the dependence of the damping of acoustic waves on the gas composition and pressure.
If, as a consequence of acoustic resonances, the discharge path is partly dis-
placed toward the wall of the discharge tube or is partly constricted, this
could result in the tube cracking due to local overheating, in instabilities with
undesirable light fluctuations, or in lamp-voltage increase. It is of interest,
therefore, to know in which frequency bands these acoustic resonances occur
and in what way they can be influenced.

7.3.1 Theory of Acoustic Resonances

The frequencies at which the arc distortions occur can be correlated to the
resonance frequencies of standing acoustic pressure waves. Such resonance
phenomena can also occur in a cavity (such as an organ pipe) having the
same dimensions as those of the discharge tube. In the discharge tube, these
pressure oscillations, superimposed on the average gas pressure, are produced
by modulations of the local power input. This causes travelling pressure
waves which, reflecting on the discharge tube wall, produce these standing
pressure waves at certain frequencies. This may lead to visible arc distortions.
Free, periodic pressure oscillations in gases behave according to the wave
equation (Morse, 1948; Redwood, 1960)

$$\nabla^2 p = \frac{1}{c_s^2} \frac{\partial^2 p}{\partial t^2} \tag{7.3}$$

where c_s = velocity of sound
        p = gas pressure
        t = time

The boundary condition follows from the requirement that the particle velocity perpendicular to the wall (v_n) must be zero at the wall

$$v_n = \text{grad}_n \, p = 0$$

For an ideal gas the velocity of sound can be written as

$$c_s = \left(\frac{c_p}{c_v} \frac{R_g T}{M_g} \right)^{1/2} \qquad (7.4)$$

where c_p, c_v = the specific heat capacities for monoatomic gases at con-
                        stant pressure and volume respectively
            R_g = molar gas constant
            M_g = molar mass of the gas
            T = temperature

In the following it is assumed throughout that the radially dependent temperature in the discharge can be replaced by an average temperature; the velocity
of sound does not then depend on the radial coordinate.

The appropriate solution of the wave equation in cylindrical coordinates r,
φ and z is (Morse, 1948)

$$p \sim \cos(l\varphi) \cos \left(\frac{\omega_z}{c_s} z \right) J_1 \left(\frac{\omega_r}{c_s} r \right) e^{-i\omega t} \qquad (7.5)$$

with l integer and with $\omega^2 = \omega_z^2 + \omega_r^2$.

Eq. (7.5) satisfies the requirement $\dfrac{dp}{dz} = 0$ at $z = 0$. The walls of the cylindrical
vessel are at coordinates $z = 0$, $z = L$ and $r = R$. Applying the boundary
condition to the other walls leads to

$$\omega_z = \frac{\pi n c_s}{L} \qquad \text{(n integer)} \qquad (7.6)$$

and

$$\omega_r = \frac{\alpha_{lm} c_s}{R} \qquad \text{(m integer)} \qquad (7.7)$$

Here α_{lm} is the mth zero of J_1', which is the first derivative of the Bessel function
J_1 of order l. Values of α_{lm} are listed by Abramowitz and Stegun (1965). Combining Eqs (7.6) and (7.7) gives the acoustic resonance frequencies in a
cylindrical tube

$$v^2_{lmn} = \left(\frac{\alpha_{lm} c_s}{2\pi R} \right)^2 + \left(\frac{n c_s}{2\pi L} \right)^2 \text{ with } n = 0,1,2,.. \qquad (7.8)$$

where $v = \dfrac{\omega}{2\pi}$

For increasing values of α_{lm}, the interseparation between successive values
of α_{lm} decreases (Abramowitz and Stegun, 1965). This implies that at increas-

ing frequency, the spectrum of resonance frequencies approaches a continuum, while at low frequencies the resonant frequencies are more widely spaced. It also appears that there is no upper limit to the α_{lm} values. So if there were no damping processes and the excitation energy were still sufficient at higher frequencies, acoustic resonances would occur at any frequency in the very high frequency region.

Apart from the factors α_{lm} and π, the resonant frequencies are determined by the inner dimensions of the discharge tube and the velocity of sound. Because the acoustic wave is reflected against the ends of the tube rather than against the tips of the electrodes – unless the electrodes fill a large part of the discharge tube – the inner discharge tube length is relevant for calculating acoustic resonance frequencies. The velocity of sound is largely determined by the average molecular weight of the gas. In standard HPS lamps, the average molecular weight depends on the partial pressures of sodium, mercury and xenon. In lamps with high mercury or xenon pressure the average molecular weight is approximately that of mercury or xenon.

A good fit with the measured resonance regions is obtained when a velocity of sound for longitudinal resonances (l and m = 0) is taken that is different from that for azimuthal-radial resonances (l or m = 0); for the longitudinal resonances in the 250 W HPS lamp a velocity of 450 m s^{-1} should be taken, and for the azimuthal-radial resonances 500 m s^{-1}. The different sound velocities for different resonance modes suggest that different effective temperatures are involved (see also Eq. 7.4). Some resonance frequencies as calculated for 150 W HPS lamps, are indicated in figure 7.16. These were obtained from Eq. 7.8.

7.3.2 Excitation and Damping of Acoustic Resonances

From a comparison of the measured resonance frequency regions of two 150 W HPS lamps (figure 7.16) it can be concluded that the number of modes observed in geometrically identical lamps having different buffer gas pressures, increases at higher pressures. This can be explained if one extends the wave equation for free oscillations (Eq. 7.3) by incorporating both the damping and excitation terms (Ingard, 1966; Ingard and Schulz, 1967; Schäfer and Stormberg, 1982)

$$\frac{\partial^2 p}{\partial t^2} + \Gamma \frac{\partial p}{\partial t} - c_s^2 \nabla^2 p = (\frac{c_p}{c_v} - 1) \frac{\partial P_V}{\partial t} \tag{7.9}$$

where Γ = damping coefficient

P_V = power input per unit volume of the gas associated with driving the acoustic mode (assumed to be the difference between the electrical input power and the losses by radiation and conduction)

221

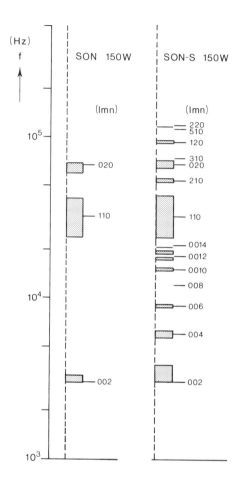

Figure 7.16 Measured current frequency bands at which resonance occur in 150 W HPS lamps with low (SON) and high (SON-S) xenon gas pressure. Hatched areas: observed resonance regions
Marked lines: calculated frequencies for several resonance modes.
The letters (l, m, n) indicate the mode of resonance occurring at the calculated current frequency.

Excitation of Acoustic Resonances

The generation of acoustic waves of the predicted resonance frequencies will only occur if the driving force of the acoustic waves is sufficiently high in the relevant frequency regions. Visible distortions by acoustic resonances occur if the input power, with a frequency in the vicinity of a resonance, is above a certain threshold value. This threshold value can be determined by supplying the lamp with a (commuted) d.c. supply current on which a high-frequency (h.f.) current is superimposed. The threshold value expressed in terms of the minimum modulation depth of the input power is given in figure 7.17 as a function of the current frequency – in this case the power frequency equals the current frequency – for a 70 W HPS lamp. If the modulation depth is less than 20 per cent, no modes are visible. However, the presence of acoustic resonances can be measured well below the threshold value for visible distortions using optogalvanic detection techniques as shown by Jongerius *et al.* (1984).

Obviously, because of the occurrence of higher harmonics, the current waveform at high-frequency operation can also be an important factor in the excitation of acoustic resonances in HPS lamps.

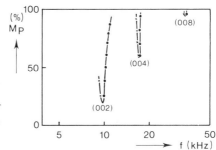

Figure 7.17 Measured minimum modulation depth of the input power causing visual arc distortion in a 70 W HPS lamp as a function of the frequency of the current superimposed on the d.c. current.

Damping of Acoustic Resonances

Damping of acoustic waves is caused by wall or volume effects. In both cases the damping coefficient increases with increasing viscosity and decreases with increasing pressure. It is likely that the variation of the damping coefficient with pressure explains the difference in the number of resonance modes observed in geometrically identical lamps with different gas pressures. One may also expect that for geometrically identical discharges the highest frequency will be observed in that having the highest gas pressure (figure 7.16). The gas pressure will have no influence on the frequency of the acoustic resonance frequencies but only on the damping term. Therefore there will only be a shift of the resonances during the warming-up of the lamp due to the changing plasma temperature and composition.

7.3.3 Correlation between Pressure Distribution and Discharge Path

In a resonant condition, the relative amplitude of the standing pressure wave can be calculated at each point in the tube by solving the wave equation (Eq. 7.5). The lines of constant pressure amplitude in a plane perpendicular to and passing through the z-axis are plotted in figure 7.18 for radial longitudinal and azimuthal longitudinal modes found in a 250 W HPS lamp (see figure 7.15). The numbers on the lines indicate their relative pressure deviations from the average value. The dashed lines indicate the nodal planes or lines. The solid lines indicate positive pressure deviations at a particular moment in time, while the dotted lines indicate negative pressure deviations. The entire pressure deviation fields in the regions of positive pressure deviations are all in the same phase with respect to the supplied power, as well as those in the regions of negative pressure deviations. The pressure distribu-

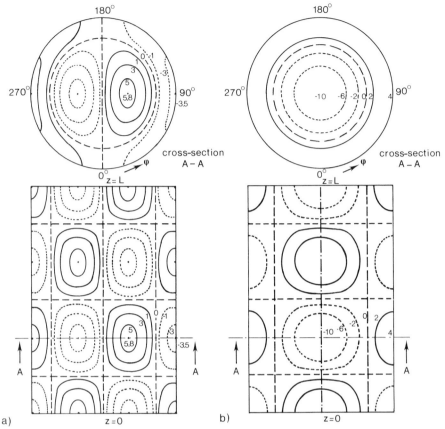

Figure 7.18 Calculated pressure distribution for the (a) (113), (b) (023) and (c) (123) modes. The lines of constant pressure amplitude are given in a plane perpendicular to the z-axis and also in a plane through this axis. The numbers on the lines indicate their relative pressure amplitudes.

tions given in figure 7.18 are all for $n = 3$ (Eq. 7.5); the extension to other values of n is obviously obtained from a repetition of the patterns in the axial direction.

Acoustic resonances are driven by the periodic power input. Changes in the power input lead to corresponding changes in the acoustic pressure according to Eq. (7.9). In a resonance situation there should be a constant phase difference everywhere in the tube between the driving input power and the resulting standing pressure wave. The discharge should choose such a path in the tube that this constant phase relation is maintained. Indeed, the calculated pressure distribution in a (113) mode shown in figure 7.18a explains the sinusoidal

224

Plate 5 (a) High-pressure sodium floodlighting of the work island 'Neeltje Jans', where work goes on round the clock in building the flood defence at the mouth of the Oosterschelde, in The Netherlands. (ILR, 1981)
(b) Beds of chrysanthemums in a greenhouse complex (Hensbroek, The Netherlands) lighted with 400 W HPS lamps, (Stoer, 1983)

Plate 6 (a) Photograph of a 70 W HPS lamp with a built-in glow-switch starter in series with a bimetal switch to put the glow switch out of circuit at high temperatures. The photograph was taken while the bimetals were being heated by the glow discharge immediately after switch-on.

Plate 6 (b) Photograph of a sodium discharge operated on a d.c. current. The sodium is transported toward the cathode side of the tube by cataphoretic effects.

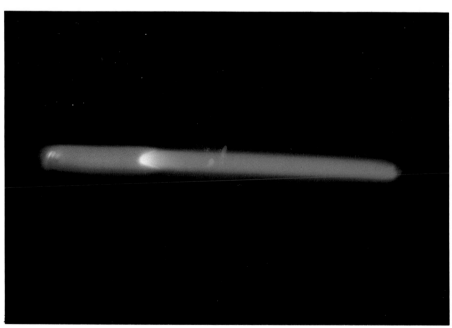

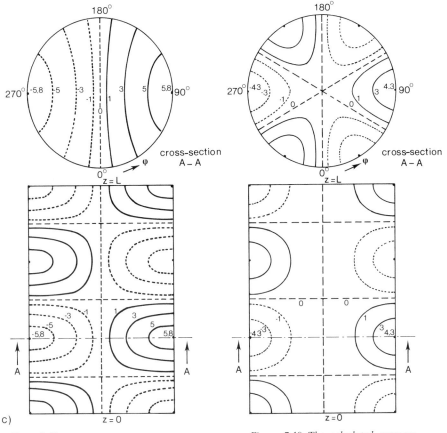

Figure 7.18c

Figure 7.19 The calculated pressure distribution of the (313) mode, not observed in HPS lamps. The lines of constant pressure amplitude are given in a plane perpendicular to the z-axis and also in a plane through this axis. The numbers on the lines indicate their relative pressure amplitudes.

discharge path of (11n) mode in figure 7.15. Also, the alternately constricted and diffuse path of the discharge belonging to the radial-longitudinal standing wave of the (02n) mode, and the sinusoidal path to one side of the tube belonging to the (12n) mode given in figure 7.15 are qualitatively in agreement with the calculated pressure fields of figure 7.18.

As an example of modes that are not observed, the calculated pressure distribution of the (313) mode in the 250 W HPS lamp is given in figure 7.19. The reason why at some frequencies the discharge path is undisturbed may be that in the pressure field there is a 'saddle' on the axis of the tube. So if the discharge is stable by itself, there is no net force to drive the discharge

225

into a curved or distorted path. The observation that modes with a saddle on the axis of the tube do not disturb the path of the discharge is in agreement with the observations of Strickler and Stewart (1963).

Disturbances of the discharge path by longitudinal acoustic resonances cannot be explained with the help of pressure distributions, because a constant phase relation cannot be found.

In conclusion, it can be said that the occurrence of acoustic resonances with visible arc distortions can be described rather well theoretically. The designer of HPS lamps destined for high-frequency or pulse operation can avoid the troublesome acoustic resonances by a proper choice of the discharge tube dimensions, gas composition and pressure. However, the higher the gas pressure chosen, the smaller are the resonance-free bands and the higher the frequency below which acoustic resonances can be excited. The ballast designer should therefore carefully choose the supply frequency when employing electronic control gear in combination with existing HPS discharge lamps.

Chapter 8

Discharge-Tube Material and Ceramic-to-Metal Seal

The discharge-tube material and the gas-tight seal between the ceramic discharge tube and the metal current feedthrough play an essential role in determining various properties of the HPS lamp with its hot, chemically aggressive sodium vapour. As already described in the Introduction (Secs 1.1.1 and 1.1.2) the development of a light-transmitting, sodium-resistant discharge-tube material was one of the crucial points for the realisation of an HPS lamp. The material properties of the discharge-tube material (Sec. 8.1) and the ceramic-to-metal seal (Sec. 8.2), as well as the manufacturing techniques employed, together ensure that the completed lamp has an optimum performance with respect to lifetime, luminous efficacy, etc.

8.1 Discharge-Tube Material

The discharge tube of the HPS lamp must meet the following requirements:
- have a high light transmittance
- be gas-tight
- be resistant to the aggressive sodium vapour for vapour pressures between 5 kPa and 100 kPa and for temperatures between 1000 K and 2000 K
- have a high electrical resistivity at these temperatures
- have a low evaporation rate at these temperatures
- possess sufficient mechanical strength and resistance to thermal shocks when the lamp is switched on and off.

Sec. 8.1.1 describes the properties of polycrystalline alumina and the process of making it; Sec. 8.1.2 deals with alternative ceramics and sapphire. Sec. 8.1.3 gives a review of the optical properties of alumina, and Sec. 8.1.4 discusses other physical and chemical properties of this material.

8.1.1 Polycrystalline Alumina

Sintered alumina, the most important properties of which are summarised in table 8.1, met all the requirements for a discharge-tube material listed above, except that of a high light transmittance. Then, between 1955 and

Table 8.1 Properties of densely sintered polycrystalline alumina (PCA). Unless stated otherwise the data are taken from Gitzen (1970) or Goldsmith et al. (1961)

Constitution	alpha alumina (α-Al_2O_3)
Crystal structure	hexagonal
Melting point	2324 K (2051 °C)
Theoretical density	3986 kg m^{-3}
Expansion coefficient at 1200 K	8.1 10^{-6} K^{-1} (polycrystalline)
	7.8 10^{-6} K^{-1} (perpendicular to c-axis)
	8.7 10^{-6} K^{-1} (parallel to c-axis)
Thermal conductivity* (1500 K)	12 W m^{-1} K^{-1}
Specific heat (1500 K)	1.31 10^3 J kg^{-1} K^{-1}
Resistivity (1000 K)	10^8 – 10^{10} Ω m
(1500 K)	10^6 – 10^8 Ω m
Ordinary refractive index at 590 nm	1.768

* After Forman (1973).

1957, a process was found for sintering alumina densely enough for it to become translucent. This was done by adding 0.2 per cent by weight of magnesia (MgO) to pure Al_2O_3 (Cahoon and Cristensen, 1955; Coble and Burke, 1961; Coble, 1962).

The process of making polycrystalline alumina involves three steps. First comes the production of a very pure (99.99 per cent) alumina powder. This can be made by calcination of alum (Al, $NH_4(SO_4)_2$ aq), giving 5 per cent γ-Al_2O_3 with a mean particle size of 0.02 μm and 95 per cent α-Al_2O_3 with a mean particle size of 0.5 μm. Magnesia is then added to sinter the alumina to translucency.

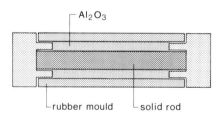

Figure 8.1 Schematic representation of the pressing process for alumina powder, in this case for a cylindrical tube (see figure 8.2a).

The second step is to give the alumina (mixed with a binder) the desired geometrical shape. This can be done, for example, by compressing the alumina powder isostatically in the annular space between a hollow, cylindrical rubber mould and a solid, cylindrical rod (figure 8.1) to produce a single cylindrical tube. Alternatively, the alumina, in the form of a paste with binder, can be extruded, or forced through a circular opening, the continuously-emerging cylinder then being cut into tubes of the desired length. This extrusion process is by far the faster of the two processes described. Non-cylindrical discharge tubes can be formed if, in the isostatic pressing process, a dummy of low-melting material (wax, tin or wood metal) in the desired shape is used instead of the cylindrical rod (Carlson, 1975). The dummy is afterwards removed by heating. This is the so-called lost-wax process. Another way of obtaining complex-shaped tubes (figure 10.21) is to employ a blowing technique to expand the extruded cylindrical tube into the shape of a surrounding cavity, a fluid pumped under pressure into the cylindrical tube being used for this purpose (Furuta *et al.*, 1983).

The third step is the sintering process. After pre-sintering the tubes in air at about 1550 K to remove the binder, the final sintering, lasting several hours, takes place in hydrogen at about 2100 K. During the sintering process, the density of the alumina increases and the tube shrinks to about 70 per

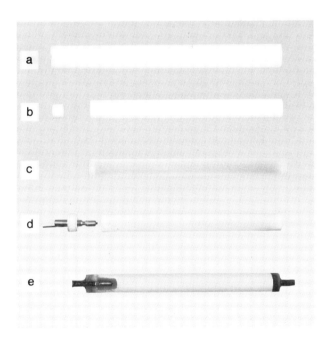

Figure 8.2 Various stages in the manufacture of a polycrystalline alumina (PCA) discharge tube for a 400 W HPS lamp:
a) cylindrical tube formed by pressing the alumina powder
b) tube and plug after pre-sintering
c) tube after the complete sintering process
d) alumina tube and electrode with its feedthrough construction
e) complete discharge tube

cent of its former size in every direction (compare tubes a and c in figure 8.2). In this way a polycrystalline material with a crystal size between 10 and 100 μm is obtained. The shrinkage can be used to sinter an electrode-feedthrough plug into place at the end of the discharge tube (tubes b and c in figure 8.2).

The density of the alumina after the sintering process is in excess of 99.5 per cent of its x-ray density (theoretical value as calculated from lattice parameters), which means that the relative volume of the pores is less than 0.5 per cent. In the process of making densely-sintered polycrystalline alumina (PCA), the addition of the correct amount of magnesia plays an essential role as it inhibits discontinuous grain growth and increases the density near to its theoretical limit (figure 8.3). The result is that pores are not trapped inside the grains (figure 8.4), but can disappear by diffusion of vacancies. So a material is produced that transmits light reasonably well, although very diffusely (Peelen, 1977).

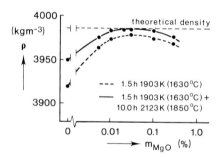

Figure 8.3 The density ρ of the sintered alumina as a function of the magnesia content m_{MgO} (given in weight per cent) for two sintering methods. (Peelen, 1977)

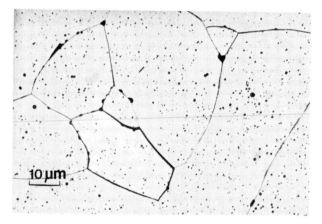

Figure 8.4 Undoped alumina sintered for 10 hours at 2123 K (1850 °C), showing discontinuous grain growth. Most of the pores are trapped inside the grains; they are visible in the photograph as black dots. (Peelen, 1977)

230

PCA scatters the light passing through it, and so, in contradistinction with sapphire, which is transparent, this material is translucent (see figures 8.5 and 8.6). Names for the polycrystalline alumina are derived from this feature: Lucalox (Registered Trade Mark of General Electric – USA) from trans-lucent aluminium oxide, and DGA from the Dutch words *doorschijnend gas-dicht aluminium oxide*, meaning translucent gas-tight aluminium oxide.

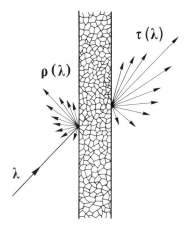

Figure 8.5 Schematic representation of the light scattering produced by a slice of porous polycrystalline alumina; λ indicates the incident radiation, τ(λ) the transmitted radiation and ρ(λ) the reflected radiation.

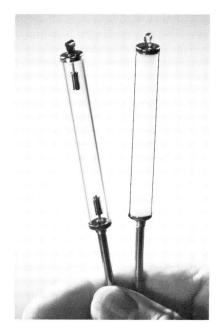

Figure 8.6 Photograph of (left) a sapphire and (right) a polycrystalline discharge tube, show-ing the difference between transparent and translucent materials. The electrodes are clear-ly visible through the transparent sapphire but not through the translucent polycrystalline alu-mina. (Rigden, 1965)

The light scattering is mainly the result of the residual porosity. This was shown by Peelen (1976, 1977), who employed another technique to sinter the alumina, namely hot-pressing. A high pressure is employed to deliver an extra driving force for sintering, and this allows the required density to be obtained at a lower temperature than is possible in the normal sintering process. The light scattering properties of hot-pressed alumina are quite different from those of normally-sintered alumina. The hot-pressed material can be described as transparent, while the normally-sintered material is translucent (see figure 8.7). The essential difference between the two materials is that the former has a much smaller mean grain size and pore size. The residual pores in the hot-pressed material have a diameter of about 0.2 μm; this is small compared to the wavelength of light. The pores in the normally-sintered alumina are about equal in size to the wavelength of light, namely around 1 μm and therefore they cause the strong scattering.

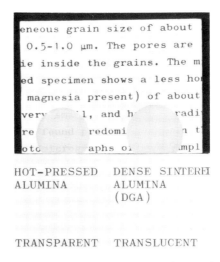

HOT-PRESSED   DENSE SINTERED
ALUMINA   ALUMINA
               (DGA)

TRANSPARENT   TRANSLUCENT

Figure 8.7 Photograph showing two discs of polycrystalline alumina (hot-pressed and normally sintered) about 10 cm above a printed page. Only the text under the hot-pressed material is legible. (Peelen, 1976 and 1977)

Dopes are added to the alumina to obtain a very high density and a regular microstructure, so improving the optical and mechanical properties of the sintered material. Besides MgO, also CaO or Y_2O_3 and La_2O_3 may be added as extra dopes (Kobayashi and Kaneno, 1975; Prud'homme van Reine, 1983). Both dopes (added foreign compounds) and impurities (foreign compounds not intentionally added) may influence the material properties. Figure 8.8, for example, shows how a CaO dope alters the microstructure of the alumina. The dopes partly evaporate during the sintering process, the dope concentration decreasing with increase in applied temperature, particularly near to the surface. The dopes and impurities may be present in solid solution, segregated at the grain boundaries, and in second-phase particles.

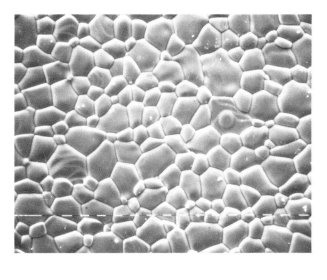

a)

——— 40 μm

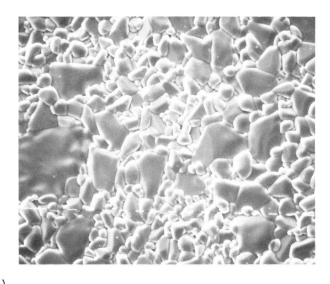

b)

Figure 8.8 Polycrystalline alumina (a) without and (b) with CaO as extra dope, apart from the MgO. In the case of the extra CaO dope, a more irregular microstructure is visible.

233

8.1.2 Alternative Ceramics and Sapphire

Of the various translucent ceramics, spinel ($MgAl_2O_4$) and yttria (Y_2O_3) may possibly serve as alternatives to alumina for use as discharge-tube material. These have higher melting points and would appear to offer optical advantages related to the cubic crystal structure (Parrott, 1974; Rhodes, 1979). For HPS lamps with an yttria discharge tube, a higher luminous efficacy has been reported (Muta and Tsukuda, 1973); this is ascribed to the higher wall temperature due to the lower emissivity of yttria compared with alumina (Waymouth and Wyner, 1981).

Discharge tubes for HPS lamps have also been made from sapphire (Labelle and Mlavsky, 1971; Loytty, 1976). The use of this monocrystalline alumina has some advantages compared with the polycrystalline material: it is more resistant to alkaline vapour (Campbell, 1972), can be operated at higher temperatures (Rickman, 1977), and the lamp's luminous efficacy is about 5 per cent higher. A disadvantage is the more anisotropic character of the monocrystalline material, the stresses in which may lead to cracks and leakages, so making sealing more difficult. This, combined with the higher price, have prevented the use of sapphire on a large scale. However, its favourable optical properties make it very suitable for use in optical studies of the HPS discharge and the electrodes.

8.1.3 Optical Properties of Al_2O_3

Transmission, Reflection and Absorption

Transmission, reflection and absorption are very important optical properties of the discharge tube material as they directly influence the luminous efficacy of the HPS lamp; they are related by the equation

$$\tau(\lambda) + \varrho(\lambda) + \alpha(\lambda) = 1 \tag{8.1}$$

where $\tau(\lambda)$ = spectral transmittance
            $\varrho(\lambda)$ = spectral reflectance
            $\alpha(\lambda)$ = spectral absorptance

The spectral transmittance and reflectance are measured with a spectrophotometer equipped with an integrating sphere (figure 8.9a and b). The intrinsic absorption of pure alumina, without impurities, is generally assumed to be small in the visible and near-infrared regions.

The scattering properties of the PCA are also important. Increased scattering can decrease the luminous efficacy of the lamp, because the radiation scattered back may be absorbed in the discharge. Because of the effects due to scattering, the in-line transmittance (see figure 8.9c for the measuring set-up)

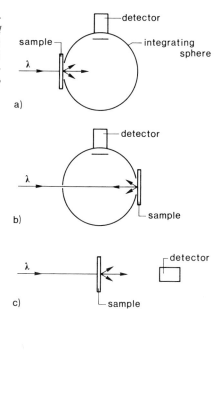

Figure 8.9 Position of sample in a spectrophotometer for measurement of (a) spectral
transmittance, (b) spectral reflectance, and (c)
spectral in-line transmittance. For methods (a)
and (b), an integrating sphere is used. The detector signal is proportional to the quantity to
be measured.

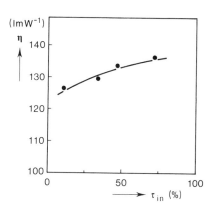

Figure 8.10 The increase in luminous efficacy
η of 400 W HPS lamps with increase in in-line
transmittance τ_{in} (in the visible wavelength region) of the alumina discharge-tube material.
(Otani and Suzuki, 1979)

of the alumina is a more sensitive measure for the quality of the discharge
tube material than is the (total) transmittance. Figure 8.10 shows the luminous efficacy of HPS lamps as a function of the in-line transmittance of the
discharge-tube material used. The spectral in-line transmittance $\tau_{in}(\lambda)$ of a
sample with thickness d is given by

$$\tau_{in}(\lambda) = [1 - \frac{2 \varrho^{\circ}(\lambda)}{1 + \varrho^{\circ}(\lambda)}] \exp[- S(\lambda) d] \qquad (8.2)$$

where $\varrho^{\circ}(\lambda)$ is the normal spectral reflectance at a single surface of the sample
and $S(\lambda)$ is the scattering coefficient (Kingery et al., 1975). The intrinsic absorption is assumed to be negligible. The normal single surface reflectance

235

is about 8 per cent for a refractive index of 1.77 in the visible region, as calculated with Fresnel's formula

$$\varrho^o(\lambda) = \left(\frac{n-1}{n+1}\right)^2 \qquad (8.3)$$

where n is the (ordinary) refractive index at the wavelength λ (Malitson, 1962).

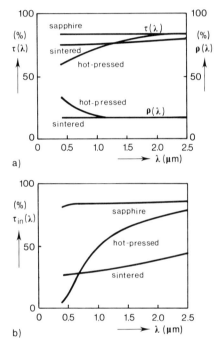

Figure 8.11 (a) The measured spectral transmittance $\tau(\lambda)$ and spectral reflectance $\varrho(\lambda)$, and (b) the spectral in-line transmittance $\tau_{in}(\lambda)$ of samples of sintered alumina, hotpressed alumina and sapphire. (Peelen, 1977)

Figure 8.11 shows the measured spectral transmittance, spectral reflectance and spectral in-line transmittance of normally-sintered alumina, hot-pressed alumina and sapphire. The large differences in in-line transmittance of these three samples are mainly due to the different scattering properties, since the spectral transmittance is more or less the same for all three samples. The difference in the scattering properties between the normally-sintered and the hot-pressed alumina is mainly due to differences in pore size ($\approx 0.2\ \mu$m for hot-pressed material and $\approx 1\ \mu$m for normally-sintered alumina), the porosity and the sample thickness being nearly the same in both cases. The scattering increases with increase in pore size, as can be shown with the theory of Mie scattering (Peelen, 1976). The surface roughness also plays a role. Chemic-

236

al polishing mainly increases the in-line transmittance and causes an increase in the luminous flux of a few per cent (Ingold and Taylor, 1982).

The quality of a PCA tube is often also assessed by determining its total light transmission. This is done by measuring the transmitted luminous flux of a miniature incandescent lamp inside the tube relative to the flux from the bare lamp (figure 8.12). The total light transmission of the tube will generally be higher than the transmittance of a flat sample from the same material.

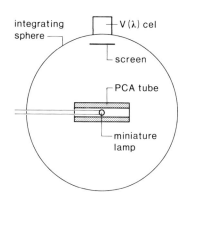

Figure 8.12 Set-up for measuring the total light transmission of a PCA discharge tube.

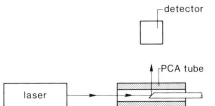

Figure 8.13 Set-up for measuring the local in-line transmittance of a single wall of a PCA discharge tube. (Jacobs, 1978)

This is because multiple reflections of light due to the alumina surface are automatically included in the measurements of the tube's total light transmission when using an integrating sphere.

The in-line transmittance of one single wall of a discharge tube can be measured locally using a laser beam, with a miniature mirror placed in the tube (figure 8.13). A similar measuring system that makes use of a laser beam perpendicular to the discharge tube, and a miniature detector inside the tube, has been described by Ingold and Taylor (1982).

Emissivity

The thermal radiation of the discharge tube, at a given temperature, depends on the emissivity of the material, which is determined by the way the tube absorbs and reflects radiation, particularly in the infrared region of the spec-

237

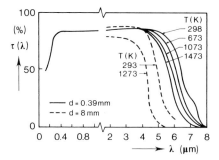
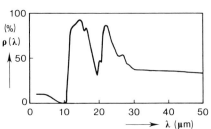

Figure 8.14 The spectral transmittance τ(λ) of 8 and 0.39 mm thick sapphire samples as a function of the wavelength at various temperatures T. (After Oppenheim and Even, 1962 and Loytty, 1976)

Figure 8.15 The spectral reflectance ρ(λ) of sapphire (single surface) as a function of the wavelength. (McCarthy, 1965)

trum. Up to about 6 μm, PCA is translucent and sapphire is transparent, the infrared transmittance being temperature dependent, see figure 8.14. In the infrared region, the reflection is strongly dependent upon wavelength. The spectral reflectance has a minimum at about 10 μm and increases rapidly above this wavelength, see figure 8.15. In general it is assumed that the reflectance does not depend on the temperature. From these transmittance and reflectance spectra, the wavelength and temperature-dependent absorption coefficient $\kappa(\lambda)$ of non-scattering materials can be calculated from the relation (Mulder and Koevoets, 1979)

$$\kappa(\lambda) = -\frac{1}{d} \ln \left[\frac{[1-\varrho^\circ(\lambda)]^2}{2\,\varrho^\circ(\lambda)^2\tau^\circ(\lambda)} \left(-1 + \sqrt{1 + \left[\frac{2\,\varrho^\circ(\lambda)\,\tau^\circ(\lambda)}{(1-\varrho^\circ(\lambda))^2} \right]^2} \right) \right] \quad (8.4)$$

where $\tau^\circ(\lambda)$ is the spectral transmittance normal to the surface.

For small reflectance values Eq. (8.4) can be approximated by the following relation

$$\kappa(\lambda) \approx -\frac{1}{d} \ln \frac{\tau^\circ(\lambda)}{[1-\varrho^\circ(\lambda)]^2} \quad (8.5)$$

The absorption coefficient of sapphire is given in figure 8.16 as a function of wavelength for various temperatures.
The normal spectral emissivity $\varepsilon^\circ(\lambda)$ is given by Touloukian and DeWitt (1970) as

$$\varepsilon^\circ(\lambda) = \frac{1-\exp[-\kappa(\lambda)\,d]}{1-\varrho^\circ(\lambda)\exp[-\kappa(\lambda)d]} [1-\varrho^\circ(\lambda)] \quad (8.6)$$

For opaque materials ($\kappa(\lambda)\,d \gg 1$) the normal spectral emissivity is

$$\varepsilon^\circ(\lambda) = 1-\varrho^\circ(\lambda) \quad (8.7)$$

238

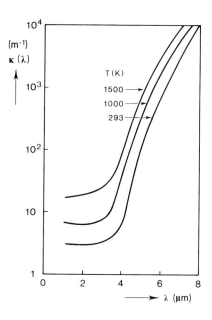

Figure 8.16 The spectral absorption coefficient $\kappa(\lambda)$ of sapphire as a function of wavelength at various temperatures. The absorption coefficient was derived with Eq. (8.4) from the transmittance and reflectance spectra given in figures 8.14 and 8.15.

For semi-transparent materials the calculation of the (hemispherical) spectral emissivity $\varepsilon(\lambda)$ – that is the ratio of the spectral radiant exitance M_λ of the material considered to the spectral radiant exitance of a black body at the same temperature – involves a complex calculation based on the radiation transport equation (Gardon, 1956). The spectral emissivity can then be calculated from the following equations

$$\varepsilon(\lambda) = 2 \int_0^{\pi/2} \varepsilon^\delta(\lambda) \sin\theta \cos\theta \, d\theta \tag{8.8}$$

$$\sin\theta = n \sin\delta \tag{8.9}$$

$$\varepsilon^\delta(\lambda) = \frac{1 - \exp\left[-\kappa(\lambda) \, d \sec\delta\right]}{1 - \varrho^\circ(\lambda) \exp\left[-\kappa(\lambda) \, d \sec\delta\right]} \left[1 - \varrho^\circ(\lambda)\right] \tag{8.10}$$

where $\varepsilon(\lambda)$ = (hemispherical) spectral emissivity
$\varepsilon^\delta(\lambda)$ = directional spectral emissivity
δ = angle with normal to surface
θ = angle of refraction or emission

For opaque materials the (hemispherical) spectral emissivity is equal to the normal spectral emissivity.

In figure 8.17 the calculated spectral radiant exitance of 1.5 mm thick sapphire at 1500 K is compared with the spectral radiant exitance of a black body at the same temperature. This figure shows that at this temperature

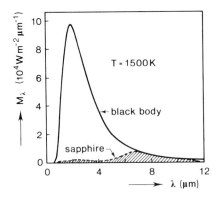

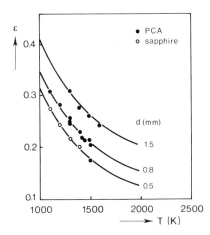

Figure 8.17 The calculated spectral radiant exitance M_λ as a function of wavelength of an 1.5 mm thick sapphire sample at 1500 K compared with that of a black body at the same temperature.

Figure 8.18 The decrease in the calculated hemispherical emissivity ε (solid lines) of sapphire tubes with increase in temperature for various values of the thickness (see text). The measured values for polycrystalline alumina (PCA) and sapphire discharge tubes with thickness d are also given. (Netten, 1975)

the thermal radiation of Al_2O_3 is much less than that of a black body radiator. The total, hemispherical emissivity ε now follows from the relation

$$\varepsilon = \int\limits_0^\infty \frac{\varepsilon_\lambda(T)\, M_\lambda(B)\, d\lambda}{\sigma\, T^4} \tag{8.11}$$

where $M_\lambda(B)$ is the spectral radiant exitance of a black body at temperature T and σ is the Stefan-Boltzmann constant.

The calculated hemispherical emissivity of sapphire tubes is given in figure 8.18 as a function of temperature for various values of the wall thickness d. For these calculations it is assumed that the thermal radiation of this cylindrical tube is equal to that of a flat sample with thickness $2d$. The calculated emissivity values as given in figure 8.18 are higher than the values given by Liu (1979), probably because of differences in the basic data employed.

The emissivity of the discharge tube can also be determined experimentally from the rate of wall temperature decay after switching off the lamp (Netten, 1975; Wyner, 1979a). In the middle of relatively long discharge tubes, where the axial temperature gradient is zero, the wall temperature as a function of time is given by the relation

$$\pi\left[(R+d)^2-R^2\right]\varrho\, c_{\mathrm{p}}\frac{\mathrm{d}T}{\mathrm{d}t} = -2\,\pi\,(R+d)\,\varepsilon\,\sigma\,T^4 \qquad (8.12)$$

where R = inner radius of discharge tube
ϱ = density of Al_2O_3
c_{p} = specific heat of wall material

This relation is valid for a discharge tube in an evacuated outer bulb, so that the losses due to conduction and convection can be neglected. Further it is assumed that thermal or reflected radiation from the outer bulb may be neglected. The emissivity values of sapphire and PCA, as determined experimentally using Eq. (8.12), are in close agreement with the calculated values for sapphire (figure 8.18). The emissivity of PCA is apparently nearly the same as that of sapphire.

The emissivity of the discharge tube determines the wall temperature for given conduction and absorption losses (Sec. 2.4) or given wall loading (Sec. 4.2).

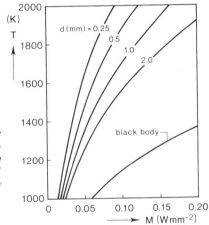

Figure 8.19 The calculated increase in temperature of a cylindrical discharge tube with increase in radiant exitance M for an alumina discharge tube with thickness d (curves are given for d = 0.25, 0.5, 1 and 2 mm). The temperature curve for a black body radiator (emissivity ε = 1) is also shown.

As an illustration of this, figure 8.19 gives the calculated temperature of alumina discharge tubes as a function of the radiant exitance (thermal radiation), compared with the calculated temperature of a black body radiator ($\varepsilon = 1$). For a given radiant exitance the wall temperature of an alumina discharge tube is higher than that of a black body radiator, in particular for small values of the wall thickness. This effect is stronger at higher temperatures, as the emissivity of alumina decreases with increasing temperature (see also Zollweg, 1983).

241

8.1.4 Other Physical and Chemical Properties of Al_2O_3

Mechanical Strength

The mechanical strength and the related thermal shock resistance of an alumina discharge tube depend on the microstructure of the alumina. Because of the anisotropic expansion of the individual crystals, the mechanical strength decreases with increasing grain size (figure 8.20). The mechanical strength needed thus imposes an upper limit on the grain size. According to Kobayashi and Kaneno (1975), the mechanical strength of PCA can be improved by the addition of dopes, for example Y_2O_3 and La_2O_3, which influence the microstructure of the alumina.

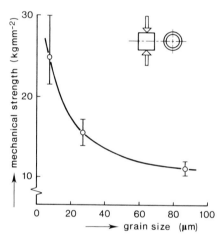

Figure 8.20 The reduction in mechanical strength of a polycrystalline alumina discharge tube with increase in grain size. The test method is indicated by the sketch in the figure. (Otani and Suzuki, 1979)

Sublimation

Sublimation of Al_2O_3 may occur at high temperatures and is mainly due to the following reactions

$$Al_2O_3 \rightarrow 2Al_{(g)} + 3O_{(g)} \tag{8.13}$$
$$Al_2O_3 \rightarrow 2AlO_{(g)} + O_{(g)} \tag{8.14}$$
$$Al_2O_3 \rightarrow Al_2O_{(g)} + 2O_{(g)} \tag{8.15}$$

where $_g$ denotes the gaseous state.

The first reaction is the most important one, as follows from the thermodynamically calculated partial pressures of the various species above alumina in vacuum (figure 8.21). The oxygen is scavenged by the barium getter in the lamp, and the aluminium forms a light-absorbing layer on the inner sur-

242

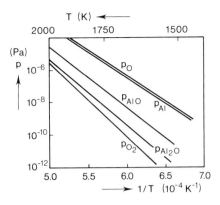

Figure 8.21 *The increase in calculated (equilibrium) partial vapour pressures p with increase in temperature for various species above Al_2O_3 in vacuum. (Campbell and Kroontje, 1980)*

face of the outer bulb, so leading to a lower luminous efficacy of the HPS lamp (Hanneman *et al.*, 1969b; Rickman, 1977; Campbell and Kroontje, 1980). The rate at which the thickness of this aluminium layer increases in an evacuated outer bulb is proportional to the aluminium pressure as given by the relation (Hanneman *et al.*, 1969b; Rickman, 1977)

$$\frac{ds}{dt} = C\frac{A_1}{A_2}\frac{1}{\varrho}\,p_{Al}\,\left(\frac{M_{Al}}{T}\right)^{1/2} \tag{8.16}$$

where s = thickness of aluminium layer
t = time
C = constant
A_1 = surface of discharge tube
A_2 = surface of outer bulb
ϱ = density of (aluminium) layer
p_{Al} = equilibrium aluminium vapour pressure
M_{Al} = aluminium molar mass
T = temperature of discharge tube

The calculated thickness of the aluminium layer as a function of wall temperature, after various burning times, is given in figure 8.22. The alumina sublimation is strongly temperature dependent, the aluminium vapour pressure increasing rapidly with temperature as shown in figure 8.21. There will be a significant darkening of the outer bulb when the thickness of the aluminium deposit is between about 50 and 200 Å. From figure 8.22 it can be seen that for a burning time of 10 000 hours the wall temperature should therefore not exceed 1500 K if darkening is to be avoided.

Sublimation can be reduced by introducing a fill-gas (10^5 Pa argon during operation) into the outer bulb, which may raise the permitted wall temperature to 1800 K as far as sublimaton is concerned (Campbell and Kroontje, 1980). The hydrogen pressure should then be extremely low ($< 10^{-4}$ Pa),

243

otherwise the reduction of Al_2O_3 by H_2 will significantly increase the alumina sublimation. The sublimation is less for sapphire tubes than it is for PCA tubes, so that for the former, higher wall temperatures are permitted than for the latter; surface irregularities may play a role for the PCA material (Rickman, 1977).

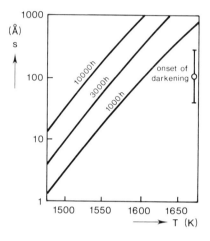

Figure 8.22 Calculated increase in thickness s of deposit on outer bulb due to sublimation from the alumina discharge tube with increase in temperature after various burning times. Darkening sets in for a deposit thickness of about 100 Å. (Hanneman et al., 1969b)

Sodium Migration

The transport of sodium through alumina under the influence of an electric field was studied by Wyner (1979b). When the outer surface of the PCA discharge tube is negatively biased, electrolysis of sodium occurs through it, and this is accompanied by black-spot formation on the tube wall. This electrolysis is probably related to the formation of black spots and the cracking observed in early high-wattage lamps (Lin and Knochel, 1974). Ignition wires round the discharge tube (Sec. 6.1.1), create high local electric fields and may also produce electrolysis. These wires are therefore usually switched off during operation, or else they are capacitively coupled to one of the electrodes with a very high 50 Hz impedance to give negligibly small electrolysis currents. Another method of avoiding electrolysis is to make the potentials of the starting electrode and the local plasma equal under operating conditions. This is done by using a voltage divider incorporating a temperature-dependent resistor (van Vliet and Broerse, 1977).

Wyner (1979b), in his experiments, found that the moment of discharge tube failure caused by sodium penetration is reached earlier with decrease in wall thickness and with increase in the temperature of the wall. Sapphire is impervious to sodium penetration, and in tests with sapphire discharge tubes no black spot formation is found. This indicates that in PCA the sodium diffusion takes place mainly along the grain boundaries.

Similar studies of sodium migration at high and low field strengths (about 250 V mm^{-1} and 80 V mm^{-1} respectively) were performed by Prud'homme van Reine (1981, 1982) in which the damage to the PCA discharge tube was studied using electron microscopy and microprobe analysis. It appears that the mechanism of the sodium migration depends on the value of the field strength employed. At high field strengths a current breakdown through the PCA may occur, producing a track of molten material along the grain boundaries (figure 8.23). At low field strengths sodium penetration along grain

outer side

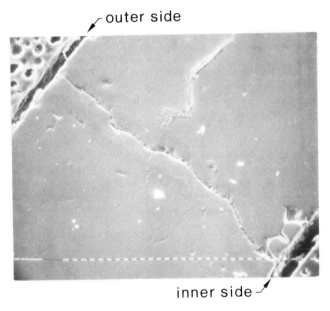

inner side

Figure 8.23 Scanning electron microscope (SEM) picture of a polished cross-section of a PCA tube at a field strength of 250 V mm^{-1}. The PCA is damaged on the inner side of the tube, showing micro-cracks along the grain boundaries. There is also a continuous crack running through the wall of the tube near to the negative electrode. (Prud'homme van Reine, 1981)

boundaries of the PCA, accompanied by chemical reactions and by micro-cracking, occurs at the inner surface of the discharge tube (figure 8.24). Sodium ions gradually migrate to the outer surface of the discharge tube, and this may eventually lead to lamp leakage. The speed of sodium migration depends strongly on the concentration of dopes and impurities, the products of their interactions with the alumina (segregating on grain boundaries in the PCA, e.g. CaO.6Al$_2$O$_3$) doing most to accelerate the sodium migration.

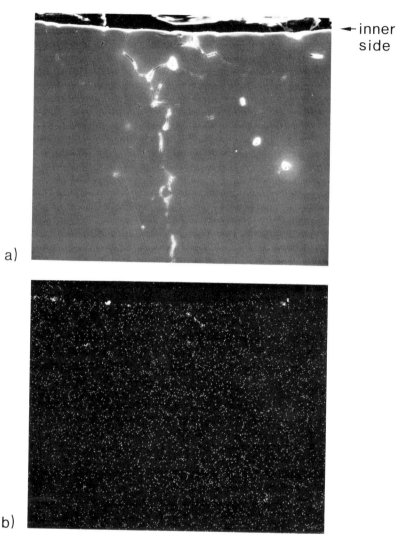

← inner side

Figure 8.24 (a) SEM picture of the polished cross-section of a PCA tube at a field strength of 80 V mm⁻¹, and (b) Na mapping of microprobe analysis showing the sodium distribution as white dots in the same cross-section. The sodium migration occurs mainly along the grain boundaries. (Prud'homme van Reine, 1981)

Attack of Discharge-Tube Material

Although it is generally assumed that the alumina is resistant to sodium vapour, this is no longer true for HPS lamps at very high sodium vapour pressures and very high wall temperatures. In this case, complex chemical reactions take place between the discharge-tube material and the sodium or other

246

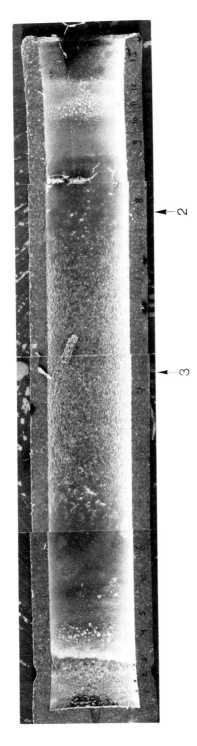

Figure 8.25 SEM picture of the axial cross-section of a 30 W HPS discharge tube after 1350 hours of operation at a relatively high sodium vapour pressure. In the middle of the discharge tube, needle-shaped crystals of sodium aluminate are visible while at the ends there are black deposits of Al, Na, Ca, Ba, W and Nb. Radial cross-sections at the positions indicated by the numbers 2 and 3 are given in figure 8.26. (Prud'homme van Reine, 1982)

components of the lamp filling (see e.g. the very recent Proceedings of the Symposium on Science and Technology of High Temperature Light Sources: Datta and Grossman (1985), Hodge (1985), Ishigami (1985), Luthra (1985), Tierman and Shinn (1985), Vrugt (1985)).

The possible consequences for the wall of the discharge tube are illustrated in figure 8.25, which shows the inner surface of an HPS lamp after 1350 burning hours at a relatively high sodium vapour pressure. Three zones may be discerned

1. A zone near the electrodes covered by a black deposit. This deposit, where elements such as Na, Al, W, Ba, Ca and Nb are present, probably comes from the sodium amalgam filling, the alumina discharge tube, the electrode, with its emitter material, the CaO containing sealing ceramic, and the niobium feedthrough, respectively.

2. A relatively clean zone between zone 1 (near the electrodes) and zone 3 (midway between the electrodes), where there is a sodium penetration along the grain boundaries (figure 8.26a and c), similar to the sodium migration discussed in the preceding paragraph.

3. A grey zone, in this case in the central part of the tube, midway between the electrodes where the wall temperature is highest. This zone has a high sodium content in the bulk of the PCA just below the inner surface, see

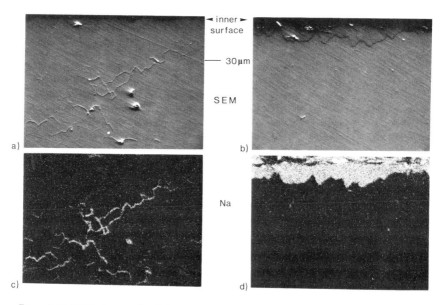

Figure 8.26 SEM pictures of radial cross-sections of a 30 W HPS discharge tube (a) 5 mm from the end and (b) half-way between the electrodes (indicated by the numbers 2 and 3 respectively in figure 8.25). Na mappings showing the sodium distribution as white dots in the same cross-sections are given in (c) and (d).

figure 8.26b and d. This sodium reacted with the alumina and is mainly present in the form of needle-shaped crystals of sodium aluminate, which rapidly grow along the grain boundaries of the PCA (see figure 8.27). The sodium aluminates are probably formed by the following type of reactions (Vrugt, 1985)

$$2\,Na + (x + \tfrac{1}{3})\,Al_2O_3 \rightarrow Na_2O.xAl_2O_3 + \tfrac{2}{3}Al \tag{8.17}$$

where $x = 1$, $x \approx 5$ and $x \approx 11$ in the case of sodium mono-aluminate, sodium β''-alumina and sodium β-alumina, respectively.

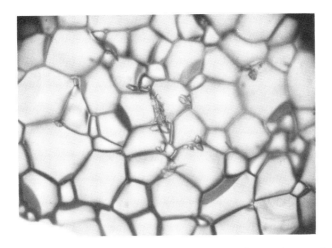

Figure 8.27 Photograph of the inner wall of a white HPS lamp showing the crystals of sodium aluminate (probably $Na_2O.11\,Al_2O_3$) at the surface. (Prud'homme van Reine, 1982)

The occurrence of Al emission lines in the spectrum gives direct evidence for the production of atomic Al (Reints Bok, 1985). Traces of oxygen present in the lamp also play a role as sodium, oxygen and alumina may react to give $Na_2O.11Al_2O_3$ and $NaAlO_2$ (Luthra, 1985). The amount of sodium involved in all these reactions, increases with increasing sodium vapour pressure and increasing wall temperature, and is thus greatly dependent upon the wall loading (figure 8.28). The amount of sodium aluminate formed by reaction (8.17) can be reduced by keeping the wall temperature below a value of 1400 K (Mizuno *et al.*, 1971; Akutsu *et al.*, 1984).

Apart from the temperature, the various reactions depend on the quality of the PCA tube, as they may be triggered or accelerated by impurities. Calcium in particular may play an important role (Hing, 1981), as is also shown in the sodium migration studies described in the preceding paragraph. CaO may destabilise the PCA, as the calcium aluminate compound $CaO.6Al_2O_3$ has a large structural similarity to the β-alumina compound $Na_2O.11Al_2O_3$ (de With *et al.*, 1985). Large amounts of MgO promote the sodium reactions. The material released from the electrodes or from the seal may also enhance the various reactions and material transports. Further, the way the alumina is sintered influences the resistance to sodium corrosion: e.g. vacuum sintering yields a more resistant material than does hydrogen sintering (de With *et al.*, 1985).

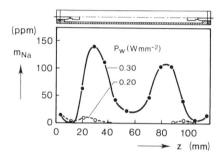

Figure 8.28 The sodium content m_{Na} (given as weight fraction) in the PCA tube wall (present as sodium aluminates) as a function of the axial coordinate z for white HPS lamps after 500 hours of operation with wall loadings of 0.20 and 0.30 W mm^{-2}. Inner diameter of discharge tube D = 7.65 mm. (Akutsu, 1984)

Due to the sodium penetration into the wall there is a loss of free sodium in the discharge tube, which leads to a changed amalgam composition. As the amalgam becomes richer in mercury, the mercury vapour pressure increases, for a given coldest spot temperature, and so produces an increase in the lamp voltage (Secs 5.6 and 10.1.2). Furthermore, the wall blackening leads to a lowering of the luminous efficacy. Strong sodium penetration into the wall may eventually result in lamp leakage. As already stated, these negative effects can be reduced by optimising the wall temperature, and by careful control of impurities in the PCA.

8.2 Ceramic-to-Metal Seal

The ceramic-to-metal seal must form a gas-tight connection between the ceramic discharge tube and the metal current-feedthrough tube or wire. The properties of the PCA tube material have been discussed in the preceding section. The metal chosen for the current feedthrough to the tungsten electrodes is usually niobium (with an addition of 1 per cent zirconium), because its coefficient of expansion is much closer to that of sintered alumina than that of tungsten itself (figure 8.29). Niobium is chemically resistant to sodium at high temperatures. Furthermore, niobium has a relatively high permeabili-

250

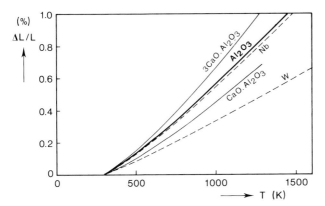

Figure 8.29 *The relative thermal expansion $\Delta L/L$ of various materials as a function of the temperature.*

Figure 8.30 *Various types of HPS discharge tube end constructions.*
All constructions use a sealing ceramic as ceramic-to-metal seal, except construction (e) where reactive metals are used. (After McVey, 1980 and Akutsu, 1984)
(a) monolithic PCA tube with alumina disc and niobium tube
(b) PCA tube with alumina disc-cap and niobium tube
(c) PCA tube with alumina disc-cap and niobium wire
(d) PCA tube with niobium cap and niobium (exhaust) tube
(e) PCA tube with niobium cap and niobium tag on the right and niobium (exhaust) tube on the left.

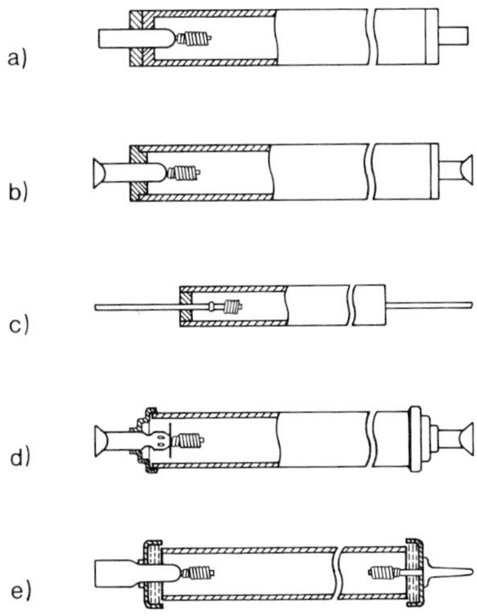

ty for hydrogen, and so hydrogen impurities in the discharge tube (produced by residual water in the discharge tube during the processing) can be gettered by the barium getter in the outer bulb (Campbell and Kroontje, 1976). The niobium current feedthrough has the form of a tube or wire, on which the tungsten electrode is generally brazed (with titanium) or welded (figure 1.6). The ends of the discharge tube are partially closed with niobium or alumina caps or with alumina discs (figure 8.30).

The gas-tight seal between the ceramic (alumina discharge tube, alumina disc) and the metal (niobium current feedthrough, niobium cap) must meet the following requirements

251

- good adhesion to the alumina
- good adhesion to the niobium
- resistant to the hot, aggressive sodium vapour at the operating temperature
- resistant to the thermal shocks that occur when the lamp is switched on and off.

Sealing materials of two types, namely reactive metals and sealing ceramics, are dealt with in Sec. 8.2.1. In Sec. 8.2.2 the sealing process is considered together with the consequences of sodium attack on the seal and with some technological aspects of the seal construction.

8.2.1 Sealing Materials

Reactive Metal Seal

A reactive-metal seal consists of thin layers of zirconium, titanium and vanadium, as indicated by the dashed lines in figure 8.30e (Rigden *et al.*, 1969). These layers are heated to about 1700 K in vacuum with a certain pressure on the end caps. At these high temperatures titanium and zirconium are capable of reducing the alumina by forming titanium and zirconium oxides, the resulting Zr-Ti-V alloy forming a seal between the niobium and the alumina.

Sealing Ceramic

The sealing ceramic is commonly based on a mixture of the oxides Al_2O_3 and CaO with additions such as MgO, BaO, B_2O_3, SiO_2, SrO or Y_2O_3 (Ross, 1966; Burggraaf and van Velzen, 1969; McVey, 1979; Chu, 1979; Otani *et al.*, 1982). The oxides Al_2O_3 and CaO are primarily chosen because of the requirements for stability, sodium resistance and low melting point. The additives are chosen to improve, amongst other things, the adhesion between the sealing ceramic and the niobium. The composition of a typical sealing ceramic is given in table 8.2. Very recently a new class of sealing materials has been developed, based on rare earth aluminates (Sc_2O_3-Y_2O_3-Ln_2O_3-Al_2O_3) with a better resistance to sodium corrosion (Oomen and Rouwendal, 1985).

Table 8.2 Typical composition of sealing ceramic for HPS lamps (type DA 1932). (Burggraaf and van Velzen, 1969)

	Mole per cent
CaO	51.4
Al_2O_3	33.1
MgO	9.5
BaO	4.2
B_2O_3	1.8

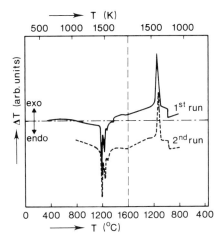

Figure 8.31 Differential thermal analysis signal ΔT as a function of temperature for the sealing ceramic DA 1932. In the first run the temperature increased from 20–1600°C (about 300–1900 K) followed by a decrease from 1600–800°C (about 1900–1100 K). In the second run the temperature ranges were 800–1600°C and 1600–800°C. Heating and cooling rate was 0.2°Cs^{-1}. (Oomen et al., 1985)

The oxide mixture – also called the sealing frit – is such that when rapidly cooled from its molten state it forms a glass, which explains the often-used name sealing glass. The thermal behaviour of a sealing ceramic is illustrated in figure 8.31 by a differential thermal analysis (DTA) curve. The endothermic dips at 1500 K, indicating large absorption of heat, show the melting of the different crystalline phases. When cooling down slowly, the crystallisation of the various phases can be traced from the exothermic peak at 1400 K.

8.2.2 Sealing Process

In a practical sealing process the sealing ceramic is applied as a ring (figure 8.32). At a temperature of 1500–1700 K (depending on the composition) the sealing ring melts and the liquid flows in the capillary-like space between the niobium and the alumina. Complex reactions and diffusion processes now take place between the sealing ceramic, the alumina and the niobium, and on cooling a solid, gas-tight layer is formed that is resistant to sodium and to thermal shocks. The results of such reactions and diffusion processes are illustrated in figure 8.33, where the continuous distribution of various elements in the seal boundaries is shown. The seal now consists of a glass phase and various crystalline phases (e.g. $3CaO.Al_2O_3$, $12CaO.7Al_2O_3$, $CaO.Al_2O_3$, $CaO.6Al_2O_3$, $MgO.Al_2O_3$), the composition depending on the rate of cooling and the starting mixture (Chu, 1979). The various crystalline phases have their own expansion coefficients (see the examples in figure 8.29). An optimum mix of such crystallites gives a seal closely matching the expansion of both the alumina and the niobium. The bond between the sealant and the alumina is better than that between the sealant and the niobium. Intermediate layers, such as silicon metal and oxides of molybdenum and tungsten, located between the niobium and the sealing ceramic, may give a better bond here (McVey, 1979; Bhalla, 1979a).

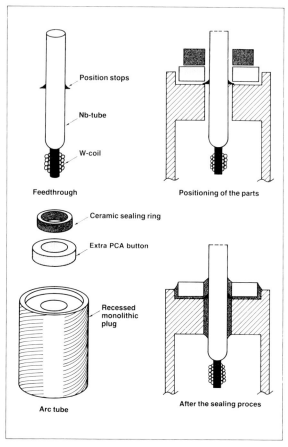

Figure 8.32 Schematic representation of the various stages in the sealing process. (Oomen et al., 1985)

Sodium Attack of Seal

The resistance of the sealing ceramic to the sodium vapour decreases with increasing sodium vapour pressure (Burggraaf and van Velzen, 1969). The photographs in figure 8.34 show, as an illustration of this phenomenon, the sodium distribution in the seal of an HPS lamp at the end of its life. From the Na and Ca distribution shown it can be concluded that there is an exchange between sodium and calcium. To reduce the chance of leakage for the HPS lamp, the seal temperature and the area of the sealed zone exposed to the sodium vapour during operation have to be kept to a minimum; any leakage path should preferably be made as long as possible. For example, in the so-called monolithic construction (figure 8.30a), the major part of the

Alkaline earth aluminate seal in the unused condition

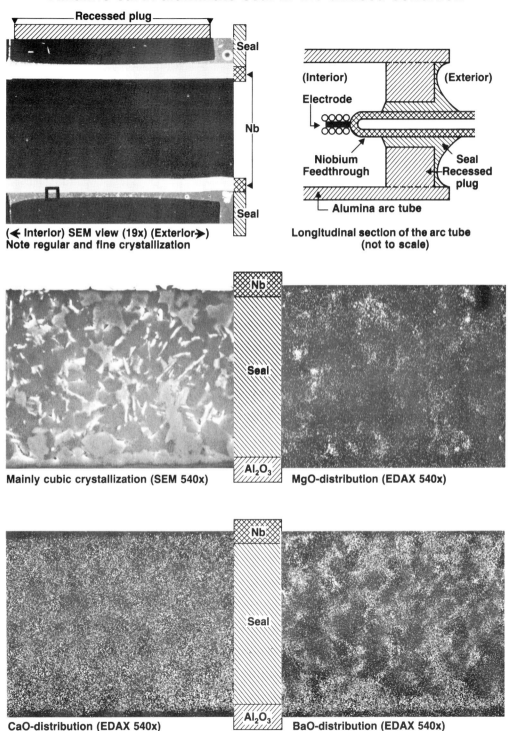

Figure 8.33 Photographs of a seal in an HPS lamp showing the crystalline phases of the sealing ceramic and the continuous distribution of Mg, Ca and Ba. (Oomen and Rouwendal, 1985)

Alkaline earth aluminate seal after 2500 hours operating at a high sodium vapour pressure (seal temperature 850 °C; P_{Na} 50 kPa)

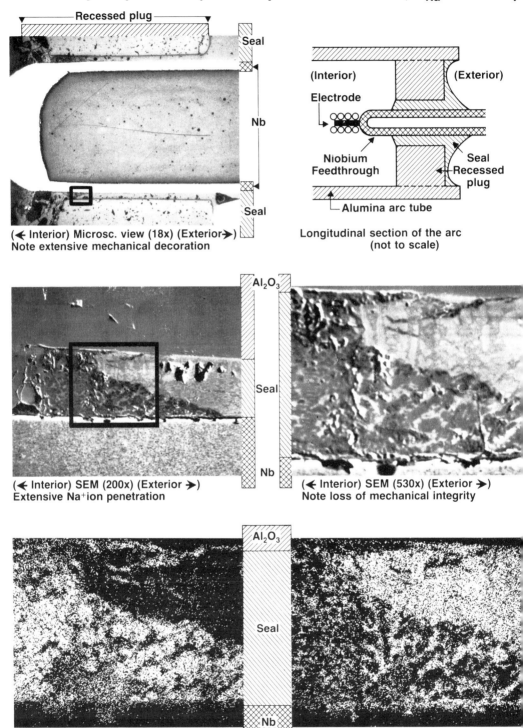

(◀ Interior) Microsc. view (18x) (Exterior▶)
Note extensive mechanical decoration

Longitudinal section of the arc
(not to scale)

(◀ Interior) SEM (200x) (Exterior ▶)
Extensive Na⁺ion penetration

(◀ Interior) SEM (530x) (Exterior ▶)
Note loss of mechanical integrity

Na₂O distribution (EDAX 530x)

CaO distribution (EDAX 530x)

Figure 8.34 Photographs of a seal in an HPS lamp after 2500 hours operating at a relatively high sodium vapour pressure. The Na and Ca distributions are shown as white dots in the photographs at the bottom. (Oomen and Rouwendal, 1985)

a)

b)

Plate 7 Shop window lighting with bowl reflector lamps: (a) incandescent lamps, (b) experimental miniature white HPS lamps. Both lamp types are used in the same type of luminaire.

HPS

HPMV

a)

HPS

HPMV

b)

Plate 8 Photographs of the discharge, terminating in the cathode phase (a) on the first winding of the coil and in the anode phase (b) on the frontside of the protruding rod. In the high-pressure sodium (HPS) discharge the arc remains diffuse up to the electrode surface. In the high-pressure mercury vapour (HPMV) discharge the arc constricts to a very small diameter in front of the electrode.

open end of the PCA tube is closed with a PCA plug sintered to the tube (de Vrijer, 1965, 1966). This results in a simple and very reliable construction (Austin, 1976).

Technological Aspects

The niobium tube is often used as the exhaust tube to evacuate the discharge tube and to introduce the gas filling and the sodium amalgam into the tube. With the monolithic construction the evacuation and the filling is carried out during the sealing process (Tol and de Vrijer, 1970). After sealing the first electrode, the sodium amalgam is introduced; the sealing of the second electrode takes place in a chamber process where the gas pressure is determined by the desired gas filling of the HPS lamp. The ceramic sealing ring is heated by placing a graphite oven around the discharge tube, which is heated using an RF generator.

8.3 Final Remarks

The various lamp constructions each have their own specific advantages and disadvantages with respect to reliability, price, etc. In the past, significant improvements in HPS lamp technology have taken place, which have led to a considerable improvement of lifetime and luminous efficacy. It may be expected that this trend will continue in the future, leading to improved materials and constructions. This is especially the case for the white HPS lamp, which imposes much higher requirements with respect to the resistance to attack by sodium than the types that are in general use at present.

Chapter 9

Electrodes

The electrodes in a gas discharge lamp act as an interface between the gas discharge and the electrical circuit. Because the HPS lamp is normally operated on an alternating current, each electrode serves in turn during one half of the period of the current supply as the cathode and as the anode. Furthermore, the electrodes have to perform their function both during ignition as well as during run-up and stable operation, i.e. under glow and arc discharge conditions, and at very different gas compositions and pressures. Consequently, the electrode design will be a delicate compromise between the very different demands of the circumstances in which the electrodes have to function. Finally, during stable operation, part of the electrical input power is dissipated in the electrodes (the electrode power) and so is lost for light generation. For HPS lamps of low power ratings in particular, the electrode losses amount to a significant part of the input power, so it is important that the electrode power is kept to a minimum consistent with other performance characteristics.

In the cathode phase the current near the electrode is sustained by electrons (liberated from the cathode) and by ions bombarding it. In the anode phase the current is sustained by electrons alone as the anode is incapable of injecting ions into the plasma. If, during the ignition phase, the electrodes are surrounded by a glow discharge, then the electrons are mainly liberated from the cathode by ion bombardment. When the arc terminates on the electrode at a spot of sufficiently high temperature the emission mechanism of electrons is field-enhanced thermionic (Sec. 9.1.1).

The glow-to-arc transition during the ignition phase (Sec. 6.1.2) and the thermionic emission of electrons from the tungsten electrodes during stable operation are enhanced by the presence of an activator, which lowers the work function of the electrode surface (Sec. 9.1.2).

Typical electrodes as used in HPS lamps are shown in figure 9.1. The electrode consists of a tungsten rod with a tungsten coil wound round it in two layers. The rod is mounted on the niobium electrical lead-in as shown in figure 8.30. Because the electrode reaches such a high temperature during operation, it is made of tungsten, which has a high melting point. The activator producing material – the so-called emitter material (Sec. 9.2) – is stored in the interstices

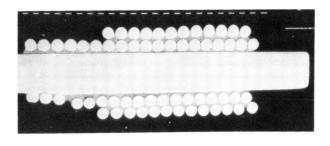

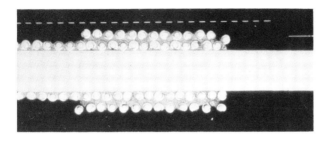

Figure 9.1 Longitudinal cross-sections of typical electrodes used in HPS lamps (a) double coil; (b) coiled-coil. The emitter material is stored in the interstices of the two layers.

of the electrode. Sputtering and evaporation effects during the ignition and stable operation phases respectively, account for the consumption of a substantial amount of the emitter material. Consequently, the emitter material is robbed of the activator, and this affects the lamp properties.

In the anode phase the diffuse discharge terminates on the entire front surface of the protruding rod; in the cathode phase it ends on a sizable portion of one of the windings of the coil. (This is very different from what happens in high-pressure mercury vapour (HPMV) discharge lamps, where the arc in front of the electrode constricts to a very small diameter – see colour plate 8.) The electrode power (mainly transferred in the anode phase to the electrode tip), is used to maintain the temperature of the electrode at the level required for thermionic emission. The electrode power is governed by the power balance of the electrode, which will be discussed in Sec. 9.3. This electrode power is lost as far as light generation is concerned, and is called the electrode losses. The measurement of the electrode power is also discussed in Sec. 9.3, and Sec. 9.4 concludes with some remarks as to the design of the electrode.

9.1 Emission Mechanism

Electron escape from a metal like tungsten is hampered by a potential barrier at the metal surface. If outside field effects at the metal surface can be neglected, electrons have to surmount this potential barrier. The energy that an electron at the Fermi-level – the upper limit of energy at temperature $T = 0\,\mathrm{K}$ – has to gain before leaving the metal, is called the work function. By heating the metal, electrons are lifted to energy levels higher than the Fermi-level and their energy is distributed according to Fermi-Dirac statistic. The higher the temperature of the metal, the more electrons in the tail of the electron energy distribution obtain sufficient energy to surmount the potential barrier. This process is called thermionic emission. The electron current density at the cathode surface is now given by the Richardson-Dushman relation

$$j_e = A\,T^2 \exp(-\frac{e\varphi_w}{kT}) \tag{9.1}$$

where j_e = electron current density
A = Richardson constant
T = temperature of the metal
e = elementary charge
φ_w = work function
k = Boltzmann constant

Electron thermionic emission is enhanced by the existence of high electric fields at the cathode and by the application of a layer of an activator on the tungsten electrode. In both cases the work function is reduced.

9.1.1 Field-Enhanced Thermionic Emission

Depending on the value of the field strength at the surface of the cathode, the following emission mechanisms can be distinguished (Dyke and Dolan, 1958)

– Field-enhanced thermionic emission caused by a lowering of the potential barrier due to the presence of an electric field – the so-called Schottky effect. The Schottky effect reduces the work function according to

$$\varphi_w = \varphi_w^o - (\frac{eE_c}{4\pi\varepsilon_o})^{1/2} \tag{9.2}$$

where φ_w^o = work function in absence of an electric field
e = elementary charge
E_c = electric field strength at the surface of the cathode
ε_o = permittivity of vacuum

The Schottky effect becomes relevant in the presence of electric fields of the order 10^5 V m^{-1}.

– Field-enhanced thermionic emission caused by electrons tunneling through the potential barrier, which is narrowed by the external field. The contribution of tunnelled electrons to electron emission becomes significant at field strengths greater than 10^8 V m^{-1}.

The electric field strength at the cathode surface can be estimated by assuming that there exists a free fall sheath for the ions in front of the cathode with a thickness of one mean-free-path length of an ion (Mackeown, 1929). The lateral extent of the cathode spot is large with respect to the sheath thickness. Therefore the field strength in the sheath can be determined from the one dimensional Poisson equation. The velocity of the heavy ions is much lower than the velocity of the electrons. Therefore, at comparable current densities of ions and electrons, the electron space charge can be neglected with respect to the ion space charge. Assuming that the field at the edge of the sheath is small with respect to the field at the cathode, the cathode field is related to the potential drop over the free fall sheath (the Mackeown equation)

$$E_c^2 \approx 4 \frac{j_i}{\varepsilon_o} \left(\frac{m_i V_f}{2e}\right)^{1/2} \tag{9.3}$$

where j_i = ion current density at the cathode
   m_i = ion mass
   V_f = potential drop over the free-fall sheath

The ion current density in the cathode spot is lower than the total current density – it is estimated to be lower than 10^7 A m^{-2} for a lamp current of 1 A and a cathode spot diameter larger than 0.3 mm (colour plate 8). The potential drop over the free-fall sheath is lower than the electrode fall values (2–15 V) as measured in HPS lamps (Sec. 9.3). The field strength at the cathode according to Eq. (9.3) is then lower than 10^8 V m^{-1}. A more detailed multi-sheath model elaborated by Waymouth (1982) predicts considerably lower field strength values of about 10^6 V m^{-1} at the cathode in HPS discharges. This means that the contribution of tunneling electrons to the emission is negligible. However the electron emission of electrodes in HPS lamps is favoured by the Schottky effect. According to Eq. (9.2), the work function may be lowered by as much as 0.4 eV, and thermionic emission will be enhanced significantly.

9.1.2 Influence of an Activator

It is well known that the potential barrier, and thus the work function of the electrode surface, can also be lowered by adsorption of a monolayer of barium atoms generated from the emitter material or by adsorption of metal atoms from plasma constituents (Waymouth, 1971, 1982). A simple, tentative explanation of the formation of monolayers of barium and sodium in HPS discharge lamps is given below (van Rijswick, 1981). The formation of a monolayer of most metals on a bare tungsten substrate is enhanced by the fact that the heat of adsorption $-\Delta H_{M/W}^{ad}$ is large as compared to the heat

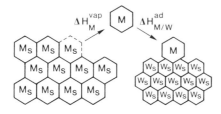

Figure 9.2 Schematic representation of the evaporation of metal atoms M_s from the bulk metal followed by adsorption on the tungsten atoms W_s.
ΔH_M^{vap} *represents the evaporation heat of the metal and* $\Delta H_{M/W}^{ad}$ *the heat of adsorption of the metal atoms on tungsten.*

of evaporation ΔH_M^{vap} of the bulk metal (see figure 9.2). According to the equation of Clapeyron, the saturation pressure $p_{M/W}$ of the metal adsorbed at a tungsten substrate is lower than the saturation pressure p_M of the bulk metal at the same temperature. At unequal temperatures of tungsten substrate T_W and bulk metal T_M, a monolayer will be formed if the vapour pressure belonging to the temperature T_M of the metal is larger than the saturation pressure $p_{M/W}$ at the operation temperature T_W of the tungsten electrode, i.e. if

$$\ln p_M (T_M) > \ln p_{M/W} (T_W) \tag{9.4a}$$

or

$$-\frac{\Delta H_M^{vap}}{R_g T_M} + \frac{\Delta S_M^{vap}}{R_g} > \frac{\Delta H_{M/W}^{ad}}{R_g T_W} + \frac{\Delta S_{M/W}^{ad}}{R_g} \tag{9.4b}$$

where ΔS_M^{vap} = change in entropy by evaporation of metal atoms from the bulk metal

         $\Delta S_{M/W}^{ad}$ = change in entropy by evaporation of atoms adsorbed at the tungsten substrate

         R_g = molar gas constant

Table 9.1 shows that $-\Delta H_{M/W}^{ad}$ is more than two times larger than $\Delta H_{M/W}^{vap}$ for the metals sodium and barium. Hence a monolayer formation will occur up to electrode temperatures that are two times higher than the temperature of the pure metal, assuming that the changes in entropy are nearly identical in both cases (figure 9.3). What the additional effect is on the formation of

Table 9.1 Measured values for heat of evaporation ΔH_M^{vap} (Stull and Prophet, 1971), heat of adsorption on tungsten $\Delta H_{M/W}^{ad}$ (Miedema and Dorleijn, 1980) and work function φ_w (Fomenko, 1966)

Metal	ΔH_M^{vap} (kJ mol^{-1})	$\Delta H_{M/W}^{ad}$ (kJ mol^{-1})	φ_w (eV)
Na	107	−246	1.8-2.1
Ba	183	−(394-467)	1.6-2.1
W			4.5

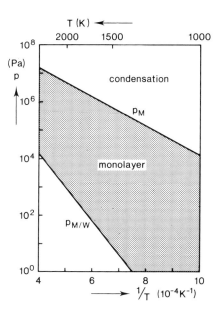

Figure 9.3 Conditions under which monolayers of sodium may be formed on electrodes. The figure shows the saturated vapour pressure of metal atoms as a function of the reciprocal of the absolute temperature. P_M is the vapour pressure in equilibrium with the metal itself and $P_{M/W}$ the vapour pressure in equilibrium with atoms adsorbed on tungsten.

The shaded area indicates the combinations of the metal vapour pressure and electrode temperature for which, under conditions of equilibrium, the tungsten surface is covered with a monolayer.

a monolayer of the barium and sodium ions returning in the electric field back to the cathode, is difficult to estimate.

Adsorption of monolayers of barium and sodium will reduce the work function of electrodes in HPS lamps (table 9.1). If barium is adsorbed by a tungsten surface covered by oxygen atoms, the work function will be lowered still more, depending on the amount of coverage with oxygen (Hermann and Wagener, 1948; Rittner et al., 1957; Rutledge and Rittner, 1957; Forman, 1979).

One questions whether, in HPS lamps, the use of emitter materials containing barium is necessary for stable operation. Especially when very high sodium vapour pressures are employed, a monolayer of sodium will probably form on the tungsten electrodes at temperatures above 2000 K, as illustrated in figure 9.3. However, for proper ignition the use of emitter materials is still found to be desirable (see Sec. 6.1.2).

9.2 Emitter Material

The purpose of the activator-producing emitter material is to reduce the work function of the electrode in such a way that ignition and stable operation of the lamp is improved. The emitter material speeds up the glow-to-arc transition during ignition and so reduces the time during which the harmful sputtering effects occur (Sec. 6.1.2). During the subsequent period of stable operation the emitter material reduces the operating temperature of the electrode, which leads to lower electrode losses and will help to avoid excessive wall blackening due to evaporation of electrode material.

In addition to maintaining a low work function the emitter material has to meet the following requirements (Bhalla, 1979b)
- if, during lamp manufacture, the emitter material is exposed to air, it should be stable at atmospheric conditions
- it should have a low evaporation rate at the electrode operating temperatures
- it should not easily be sputtered away if the emitter material is exposed to ion bombardment
- it should not interact with plasma constituents such as sodium and mercury or after evaporation or sputtering, react with the tube material or sealing ceramic.

9.2.1 Chemical Stability

The emitter material employed on the electrodes of early HPS lamps was identical to that used in HPMV lamps; for example, a mixture of $BaCO_3$, $CaCO_3$ and ThO_2 prepared by heating to a high temperature. Because of the radio-active properties of ThO_2, this oxide was replaced by Y_2O_3 (de Kok, 1975). The role of $CaCO_3$ and Y_2O_3 is to give compounds with a better stability when heating the emitter on the tungsten substrate than those obtained were only $BaCO_3$ to be used. During this heating process ('sintering') the carbonates are (at least partially) converted into tungstates by reaction with the substrate. After sintering, the electrodes may react with the air constituents H_2O and CO_2, which influences the lamp properties in a negative sense. According to Bhalla (1978, 1979b) solid solutions of niobates, tantalates, tungstates and molybdates or mixtures of these are more resistant to atmospheric contamination by H_2O and CO_2 than the sintered carbonates with Y_2O_3. Stable single-phase emitter materials such as Ba_2CaWO_6 (Smyser and Speros, 1970), Ba_3WO_6, and many other recently investigated substances (Watanabe et al., 1977; Hirayama and Bhalla, 1980) can be used in a dry environment. Another advantage of these emitter materials is that a drastic

264

shrinking of the emitter material by sintering is avoided. As shown in figure 9.4, more emitter material is then stored in the interstices of the electrodes than with the mixture of $BaCO_3$, $CaCO_3$ and Y_2O_3.

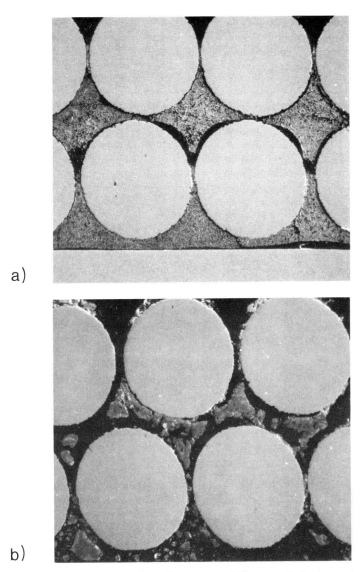

a)

b)

Figure 9.4 Contrast in quantity of stored emitter material for two typical emitter materials. Photographs show longitudinal cross-sections of the tungsten double layer of windings on two electrodes after sintering. One of the electrodes (a) was filled with $Ba_2C_aWO_6$, the other (b) with a mixture of $BaCO_3$, $CaCO_3$ and Y_2O_3. By using the tungstate, a drastic shrinking of the emitter material during sintering is avoided, so that more emitter material is stored in the interstices of the electrode than with the mixture.

9.2.2 Evaporation and Sputtering Effects

The emitter materials employed in HPS lamps are mostly tungstates containing barium. Reaction of the tungstate with the tungsten substrate generates the required free barium for the formation of a monolayer. As an example of this the most probable reaction responsible for barium release from Ba_3WO_6 is (Levitskii et al., 1979)

$$\frac{5}{3} Ba_3WO_{6(s)} + \frac{1}{3} W_{(s)} \leftrightarrows 2 Ba_2WO_{5(s)} + Ba_{(g)} \tag{9.5}$$

where s and g denote the solid and gaseous states respectively.

The weight loss of the solid emitter material Ba_3WO_6 is caused by evaporation of gaseous Ba obtained according to reaction (9.5) and of gaseous BaO produced by the reaction

$$Ba_3WO_{6(s)} \leftrightarrows Ba_2WO_{5(s)} + BaO_{(g)} \tag{9.6}$$

Under equilibrium conditions the weight losses of Ba and BaO by evaporation in vacuum are given by the evaporation equation (Langmuir, 1913)

$$G_v = 4.4 \; 10^{-6} \left(\frac{M_B}{T}\right)^{1/2} p \tag{9.7}$$

where
G_v = evaporation rate [g mm^{-2} s^{-1}]
M_B = molar mass of Ba or BaO [g mol^{-1}]
T = temperature [K]
p = pressure [Pa]

The equilibrium pressures of Ba and BaO are related to the free enthalpies ΔG_T° of the reactions (9.5) and (9.6), respectively by

$$\ln p = -\frac{\Delta G_T^\circ}{R_g T} \tag{9.8}$$

With the help of thermodynamic data obtained from literature (Levitskii et al., 1979; Chase et al., 1974) the total evaporation rate of Ba and BaO from Ba_3WO_6 and BaO in vacuum was calculated and is given in figure 9.5. The barium containing tungstate Ba_3WO_6 is preferred to emitter materials containing pure BaO (as used in low-pressure discharges) in the interest of obtaining a low evaporation rate.

During the glow discharge, which takes place while the lamp is starting, emitter material may sputter away from the electrode. Sputtering is the removal of atoms from the surface by transfer of momentum and energy from incident ions to emitter material atoms or molecules and tungsten atoms. One way of studying sputtering effects is to register, from the first moment of starting, the appearance of glow phenomena in the lamp voltage during both the positive and the negative parts of the supply-current period (see for example figure 6.18) together with the spectral distributions of radiation generated close to the surface of both electrodes (Hirayama *et al.*, 1981b; Otani *et al.*, 1981).

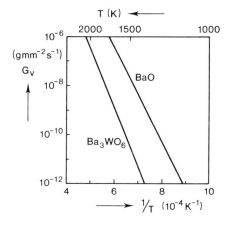

Figure 9.5 The calculated total evaporation rate of Ba and BaO together in vacuum from the emitter materials BaO and Ba_3WO_6 in thermodynamic equilibrium with tungsten.

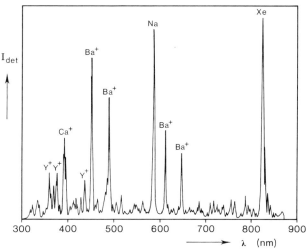

Figure 9.6 Spectrum between 300 and 870 nm of the plasma near the electrode measured within the first 0.2 s after starting of a 50 W HPS lamp with emitter coated electrodes ($BaCO_3$, $CaCO_3$ and Y_2O_3 sintered at high temperature). I_{det} is the detector signal.

In figure 9.6 such a spectral distribution, for a 50 W HPS lamp with emitter-coated electrodes ($BaCO_3$, $CaCO_3$ and Y_2O_3 sintered at high temperature) as recorded with an optical multichannel analyzer within the first 0.2 s after starting, is given by way of example. From the development in time of the intensities of the Ba, Ca and Y-ion-lines, together with the observation

267

whether or not glow phenomena occurred during the first seconds after starting, the authors came to the conclusion that the loss of emitter material during
ignition is due not only to sputtering but also to evaporation. In high wattage
HPS lamps in particular, the lamp current greatly exceeds the maximum current at which glow phenomena can still occur, so that evaporation is the
main cause of emitter material loss during starting of these lamp types. Sputtering effects occur mainly in low-wattage HPS lamps. It is for these types
that new electrode designs are proposed to minimise the ignition time (Sec.
6.1.3) within which glow phenomena occur. These electrodes are made with
an outer and an inner coil, the inner layer consisting of a thin, emitter-coated
coil emitter similar to that used in the electrodes of fluorescent lamps. The
thin tungsten wires of the inner coil and the smaller total heat capacity of
the electrode should make a faster glow-to-arc transition possible (Saito *et
al.*, 1978 – see also Sec. 6.1).

In addition to evaporation, sputtering is also a process by which emitter material is lost. Due to sputtering and evaporation of barium in the outermost
windings of the double layer the electrode becomes exhausted of this activator
material. This is shown in figure 9.7 for an electrode that has been operating

a)
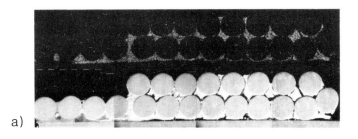

b)
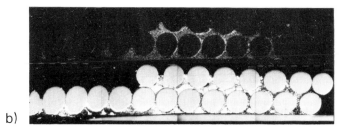

*Figure 9.7 Disappearance of Ba from the emitter material in electrodes of HPS lamps during
lamplife. The photographs were made with the aid of SEM (Scanning Electron Microscopy) and
EDAX (Energy Dispersive Analysis of X-rays). They show the electrodes of a 210 W HPS lamp
(a) after 100 h operation and (b) after 10000 h. The concentration of white dots in the upper
part of each photograph is a measure for the content of Ba as measured with EDAX. The outermost
windings of the double layer become exhausted of barium after 10000 h of operation.*

for ten thousand hours in a 210 W HPS lamp. The effect of barium depletion is that the cathode spot will move farther away from the electrode tip, so contributing indirectly to lamp voltage rise (Sec. 10.1.2).

Both evaporation and sputtering are regulated by diffusion processes near the surface of the cathode and can, therefore, be reduced by increasing the pressure of the xenon starting gas. The result of such an increase is a strong reduction of the wall blackening at the end of the discharge tube during lamp-life, as shown in figure 9.8, where two 150 W HPS lamps, one with a (cold) xenon starting gas pressure of 3 kPa and the other with a (cold) xenon starting gas pressure of 25 kPa are compared.

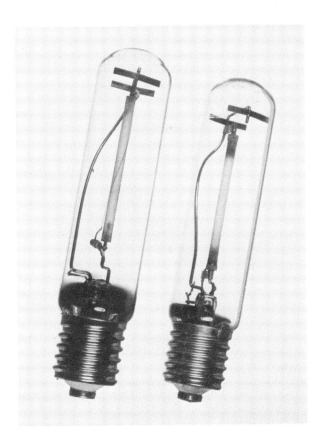

Figure 9.8 Avoiding blackening of the ends of the discharge tube by the use of xenon starting gas at high pressure. Two 150 W HPS lamps are shown, both having burned for 10000 hours. The lamp on the right, in which the xenon starting gas is at low pressure, shows blackening of the discharge tube whereas the one on the left, with high-pressure xenon, does not. The less the blackening, the better the lumen maintenance. (Jacobs and van Vliet, 1980)

269

9.3 Power Balance

From the power balance of the electrode the power necessary to maintain the required temperature for electron emission can be derived. A one-dimensional model as employed by Tielemans and Oostvogels (1983) shows how the electrode power is influenced by lamp current and electrode dimensions for a simple tungsten rod electrode. By experiments on high-pressure argon and mercury discharges they verified that an analytical solution of the one-dimensional power balance describes the relationships between electrode tip temperature, electrode power and electrode dimensions reasonably well.

9.3.1 Model

Tielemans and Oostvogels assumed that
- radial temperature gradients in the rod are negligible;
- Ohmic heating does not play a role;
- radial cooling at the sides of the rod occurs by thermal radiation only;
- the influence of ambient temperature can be neglected;
- the power is delivered by the plasma to the front of the rod-type electrode.

The power balance is then governed by the following heat transport equation

$$\frac{1}{4} \pi d_e^2 \frac{d}{dz} \kappa \frac{dT}{dz} - \pi d_e \varepsilon \, \sigma T^4 = 0 \tag{9.9}$$

where d_e = diameter of the rod-type electrode
z = axial coordinate
κ = thermal conductivity of tungsten
T = temperature
σ = Stefan Boltzmann constant
ε = emissivity of the electrode surface

By introducing the reduced axial coordinate

$$y = \frac{z}{d_e^{1/2}} \tag{9.10}$$

the heat transport equation can be rewritten as

$$\frac{d}{dy} \kappa \frac{dT}{dy} = -4 \varepsilon \sigma T^4 \tag{9.11}$$

The electrode power (also called the electrode losses) at the front of the rod ($y = 0$) is given by

$$P_{el} = I_{la} V_{el} = \frac{1}{4} \pi d_e^{3/2} \kappa \left(\frac{dT}{dy}\right)_{y=0} \tag{9.12}$$

where P_{el} = electrode power
$\quad\quad\quad I_{la}$ = lamp current
$\quad\quad\quad V_{el}$ = electrode fall

From Eqs (9.10) and (9.12) the following similarity rules for the electrode can be derived. The same temperature profiles are maintained if the electrode power, and the rod length L and diameter d_e are varied according to

$$\frac{L}{d_e^{1/2}} = \text{constant} \tag{9.13}$$

and

$$\frac{P_{el}}{d_e^{3/2}} = \text{constant} \tag{9.14}$$

If the electrode fall V_{el} does not vary strongly with the lamp current, these similarity rules can also be expressed as

$$\frac{I_{la}}{L^3} = \text{constant} \tag{9.15}$$

and

$$\frac{I_{la}}{d_e^{3/2}} = \text{constant} \tag{9.16}$$

To reduce the ignition time and to make the storage of activator material possible, other configurations than the rod type are used in practical HPS lamps (figure 9.1). In that case, part of the electrode power is dissipated during the anode phase at the electrode tip and part in the cathode phase at the first winding of the coil, as shown in colour plate 8. Furthermore, the radiating surface is enlarged by the use of the double layer of windings in which the thermal contact of the windings with the rod is mostly bad. Nevertheless, the power balance of these practical electrodes seems mainly to be determined by the rod dimensions, so the similarity rules can be used to a first approximation in determining the electrode dimensions for HPS lamps. This is shown in table 9.2, where the electrode dimensions for 50 W and 1000 W HPS lamps, as derived from the electrode dimensions of the 400 W HPS lamp with the help of the similarity rules, are compared with the actual dimensions employed in practical lamps. The small differences between calculated and actual electrode dimensions for such a large current range are due not only to the reasons mentioned above, but also to the fact that the distance between the electrode tip and the end of the discharge tube is used to regulate the coldest spot temperature, which is different for the various lamp types.

Table 9.2 Electrode dimensions (rod diameter d_e and length L) for 50 W and 1000 W HPS lamps as derived from the electrode dimensions of the 400 W HPS lamp, compared with the electrode dimensions as used in actual lamps

P_{la} (W)	I_{la} (A)	actual		derived	
		d_e (mm)	L (mm)	d_e^* (mm)	L^* (mm)
50	0.76	0.5	3.7	0.4	5.0
400	4.45	1.2	9.2	1.2	9.2
1000	10.3	1.9	12.9	2.1	12.1

9.3.2 Electrode Power Measurements

The electrode power can be determined experimentally in two ways. The first method is based on the measurement of the electrode fall V_{el} and the lamp current I_{la}, their product being assumed to represent the electrode losses. The electrode fall is then derived from the lamp voltages as measured on identical discharges differing only in discharge length according to Eq. (4.26). Denbigh and Wharmby (1976), using such a method, obtained a value of 4 ± 1 V for the electrode fall of emitter coated electrodes in HPS lamps, irrespective of electrode dimensions and lamp current. For tungsten electrodes without emitter material, an electrode fall of 8 ± 1 V was found.

The second method is based on the measurement of the temperature profile along the electrodes, from which the losses can be calculated. Such electrode temperature measurements, which were carried out by Tielemans (1974) on electrodes placed in sapphire discharge tubes, were made using a pyrometer fitted with a narrow spectral bandpass filter consisting of a monochromator. A schematic representation of this measuring system is given in figure 9.9.

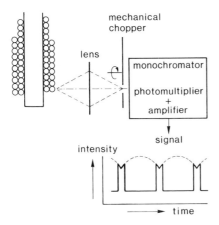

Figure 9.9 Schematic representation of the system used for the measurement of electrode temperatures.

272

The electrodes were projected on the entrance slit of the monochromator. A suitable measuring wavelength was chosen for which the plasma radiation and molecular absorption by Na_2 molecules is small (viz. $\lambda \approx 720$ nm and 890 nm). The signal was detected with the help of a photomultiplier. A mechanical chopper was employed to reduce the interfering plasma radiation intensity present at the measuring wavelength with respect to the continuum intensity originating from the electrode. To this end, the radiation was measured close to the moment of current reversal. The measuring set-up was calibrated with a special tungsten ribbon lamp (Bezemer et al., 1978). After applying a correction for the transmission of the sapphire discharge tube (see figure 8.14) and the glass outer bulb, the electrode temperature was found from the measured spectral radiance, assuming that the spectral emissivity of the tungsten electrode coated with emitter material closely approximates that of the tungsten ribbon lamp. The accuracy of the temperature measurement was assumed to be better than 25 K. The temperature at the inner end of the discharge tube – nearly equal to the coldest spot temperature – was found by determining the wavelength difference $\Delta\lambda$ between the maxima of the self-reversed D-lines (Sec. 3.2.1).

The complete temperature distribution was obtained by solving the power balance equation for the electrode as well as for the niobium feedthrough with the measured electrode and coldest spot temperature as boundary conditions. From these temperature profiles the radiation emitted by the electrode surface can be determined, as well as the power conducted away through the end of the tungsten rod. The sum of these terms represents the electrode losses.

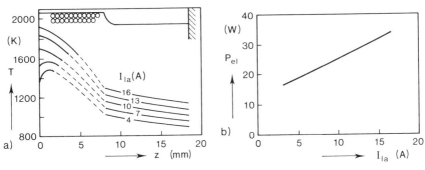

Figure 9.10 Electrode losses derived from measured electrode temperatures. (a) Measured temperature profiles on a double layer type electrode coated with an emitter, for five lamp currents. The 1.2 mm diameter rod electrodes were mounted in a sapphire discharge tube filled with only sodium, with xenon as a starting gas, (b) The electrode power P_{el} for both electrodes together, derived from the above-measured temperature profiles as a function of the lamp current I_{la}. (Tielemans, 1974)

As an example, in figure 9.10a temperature profiles are given as found for a double-layer type electrode, coated with emitter material and mounted in a sapphire discharge tube ($D = 7.6$ mm) filled with sodium. Xenon was used as the ignition gas. The electrode power or losses derived from these tempera-

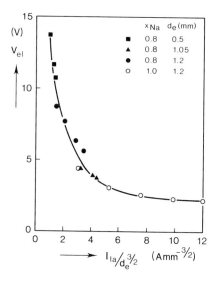

Figure 9.11 Electrode fall values (V_{el}) derived from temperature profiles as a function of $I_{la}/d_e^{3/2}$ for various electrode rod dimensions (d_e), and amalgam compositions (x_{Na} is the sodium mole fraction in the amalgam).

ture profiles are given in figure 9.10b. If, as shown in figure 9.11, the electrode fall (viz. electrode power divided by the lamp current) is plotted against the value of $I_{la}/d_e^{3/2}$ for various electrode dimensions and for different amalgam compositions the electrode fall will be found to decrease with increase in this value of $I_{la}/d_e^{3/2}$, as expected from Eq. (9.14). Thus the electrode fall cannot be approximated by a constant value if one considers a large range of $I_{la}/d_e^{3/2}$-values. In practical HPS lamps in which, in addition to sodium, mercury is used as a buffer gas, the electrode fall values are higher than in the discharges without mercury, because of the lower current values.

Although the electrode fall values found by Denbigh and Wharmby (1976) are not very different from the values given in figure 9.10 for practical $I_{la}/d^{3/2}$-values (table 9.2), a straightforward comparison with the values found from the temperature distribution is not possible because of the unknown electrode dimensions in the former case.

9.4 Final Remarks

At the end of this chapter we can conclude that the design of electrodes for HPS lamps is a compromise between the various, contradictory requirements they have to satisfy during ignition and continuous operation. Firstly, to reduce the ignition time and the electrode losses light electrodes with a thin rod and a thin coil are preferred to heavy ones. But, to prolong lamplife and improve maintenance the amount of emitter stored in the electrodes should be as large as possible because during operation the emitter is consumed by sputtering and evaporation. Moreover, the electrode design must be such that during operation a temperature distribution along the electrode is obtained that ensures the maintenance of a monolayer of barium at the cathode spot without excessive evaporation of the emitter material. Similarity rules can help the designer in scaling up or down the electrode dimensions for HPS lamps of higher or lower power ratings so that the same electrode temperature distribution is maintained.

Other points influencing the electrode design not mentioned up to now, are

- the regulation of the coldest spot temperature by the distance between the electrode rod tip and the inner discharge tube end;
- the stabilisation of the cathode spot – especially in HPS lamps with high

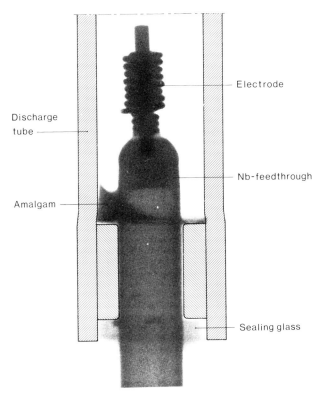

Figure 9.12 Electrical contact between amalgam and niobium feedthrough. X-ray photograph of the discharge tube end of a 70 W HPS lamp showing the amalgam, which is in electrical contact with the niobium cap so as to encourage the discharge to terminate on the former during ignition.

sodium vapour pressures – by designing an electrode with a longer protruding rod (Akutsu, 1984);

– the prevention of arcing on the amalgam during ignition by a special discharge tube end construction (Denbigh, 1983) or a greater distance between the electrode rod tip and the inner end of the discharge tube. In HPS lamps of low power ratings in particular, the arc terminates preferably on the amalgam during ignition if there is an electrical contact between the amalgam and the electrical feedthrough which is often the case in such narrow discharge tubes (figure 9.12).

Although the various factors determining the electrode behaviour in gas discharge lamps are qualitatively understood, the design of electrodes for HPS discharge lamps has, up to now, been more, a kind of art, based on experiments.

Chapter 10

Lamp Design

The design of HPS lamps in general and of its discharge tube in particular, (de Groot *et al.*, 1975; Collins and Mule, 1977; Beyer *et al.*, 1977; Iwai *et al.*, 1977; Denbigh, 1978; Denbigh *et al.*, 1983, 1984, 1985) involves trying to reach a satisfactory compromise with respect to a number of design aims, such as
– certain electrical characteristics for a given luminous flux
– long lamp life
– high luminous efficacy
– sufficiently good light quality
The relative importance of each of these design aims will depend on the particular type of application for which the lamp is to be used. For example, for public lighting a long lamp life and high luminous efficacy are more important than good light quality, which is, however, a very important design purpose of discharge lamps substituting incandescent lamps.
To obtain the best possible compromise between the several design aims, the designer is free to specify – within certain limits – such parameters as
– diameter of the discharge tube and distance between the electrodes
– sodium and mercury vapour pressures, which are determined by the choice of amalgam temperature, composition and weight
– starting gas pressure and composition – or (high) xenon buffer gas pressure
– discharge-tube wall temperature as influenced – among other things – by the wall thickness.
Other parameters, such as
– discharge-tube material
– construction of the ceramic-to-metal seal, and the material used for this
– electrode dimensions and emitter material
have been extensively discussed in Chapters 8 and 9 and can mostly be optimised independently of the first group of parameters.
The present chapter examines how each of the parameters of the first group influences the design. Frequent reference is made to data and conclusions drawn in foregoing chapters. Each of the design aims is examined in turn as a preliminary to discussing the various aspects of HPS lamp design.

10.1 Design Aims

10.1.1 Electrical Characteristics

The electrical characteristics of discharge lamps are subject to certain limitations for their ignition and stable operation.

Lamp Power, Voltage and Current

To fulfil the application requirements for public lighting mentioned in Sec. 1.2.4, the currently available lamp types cover a range of luminous fluxes from which the lamp power ratings have been derived (table 1.3). The lamp voltage cannot be chosen freely but as is discussed in Sec. 6.2 is dependent on the type of stabilisation circuit chosen. This means that in the standard case of inductive stabilisation, the lamp voltage should be about half the supply voltage. As a consequence, the lamp current values can easily be determined from lamp power and voltage, because the power factor of the lamp is nearly constant ($\alpha_{la} = 0.8 \pm 0.05$ under nominal lamp conditions).

The lamp voltage is the sum of the electrode fall and the arc voltage. By far the more important of these two terms is the arc voltage, which is the product of the electric field strength and the distance between the electrodes. The main factors influencing the field strength are the diameter of the discharge tube, the sodium vapour pressure and the mercury vapour pressure.

The sodium and mercury vapour pressures are determined by the composition, weight and temperature of the amalgam (Sec. 5.6). Even small variations in the amalgam temperature greatly influence the field strength, as is shown in figure 10.1. The amalgam temperature has therefore to be carefully adjusted, paying attention to those features of construction influencing the coldest spot temperature, such as the distance between the electrode tip and the bottom of the discharge tube.

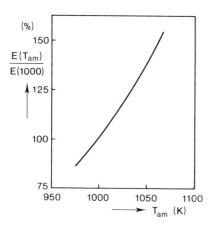

Figure 10.1 Calculated influence of amalgam temperature on the reduced electric field strength in an HPS discharge within a discharge tube of 4.8 mm inner diameter and mole fraction of sodium in the amalgam of 0.8.

Another condition is that in the standard case of inductive stabilisation at the mains frequency the reignition voltage must not rise too far above the nearly-constant value of the lamp voltage prevailing during the current-carrying part of the period – the so-called plateau voltage. In this manner premature extinction of the discharge lamp is prevented. Premature extinction occurs especially when moderate mains-voltage dips occur (figure 6.26). If the reignition voltage rises far above the plateau voltage, as is the case for low-power lamps having small discharge-tube diameters, then the lamp voltage has to be reduced to below half the supply voltage. For such lamps operating on a supply voltage of 220 V, a lamp voltage as low as 85 V may be desirable (see table 1.3).

In the case where stabilisation is achieved using a solid-state ballast (Sec. 6.2.5) the lamp voltage, and consequently the lamp current, can be chosen more freely. Also, solid-state ballasts can operate HPS lamps at frequencies where reignition peaks do not occur in the lamp voltage (figure 7.1). Thus, solid-state ballasts afford greater freedom in lamp design.

Ignition Characteristics

In general, it is true that both the luminous efficacy and the life of the HPS lamp are improved with an increase in atomic weight and pressure of the starting gas (see Chapter 5 and Sec. 10.1.2). However, breakdown is generally hampered by these measures, so that special starting aids are necessary (Sec. 6.1.1). As will be discussed in Sec. 10.2.3, different types of HPS lamps have been designed by varying the choice of starting gas composition and pressure to give the best compromise between ignition characteristics, luminous efficacy and life, to suit various areas of application.

10.1.2 Lamplife

As holds for all discharge lamps, the expression 'the life of HPS lamps is N hours' is not an unambiguous statement. The main reason for this is that lamplife is influenced by many factors, not all of which are under the control of the lamp manufacturer. Included among these factors are

- ambient temperature
- type of luminaire
- type of ballast and ignitor
- value and stability of the mains voltage
- burning position
- mode of operation (on-off switching cycle)
- severity of mechanical vibrations.

Life tests on discharge lamps performed by lamp manufacturers are therefore carried out under specified conditions.

Secondly, the term 'lamplife' as obtained from life tests with large, representative groups of lamps, can be defined in several ways. For example, the average rated life of HPS lamps is defined as the number of burning hours elapsed

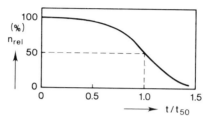

Figure 10.2 Typical survival curve for high-pressure sodium lamps. The number of survivors is given as function of burning time both in relative quantities. The average rated lamplife is reached when 50 per cent of the lamps have failed. The end of useful life occurs at a much lower mortality rate than 50 per cent. On the horizontal axis the ratio is plotted of the burning time to the average rated lamplife.

before 50 per cent of the lamps have failed (figure 10.2). A lamp fails (that is to say reaches the end of its life) when, amongst other things:
– it stops burning steadily and starts to ignite and extinguish during operation
– it fails to ignite
– discharge-tube cracking occurs
– leakage of ceramic-to-metal seal or outer bulb occurs.
Of more practical interest, however, is a knowledge of when the HPS lamp will reach the end of its 'useful life'. Useful life corresponds to a mortality rate of much lower than 50 per cent. A lamp is said to have reached the end of its useful life when it no longer provides light of the specified level (lumen maintenance) or quality (see Sec. 10.2.3 HPS Lamps with Improved Colour Rendering). Lamplife is thus determined by a number of factors, but two of the most important of these for standard HPS lamps are lamp voltage rise and lumen maintenance.

Lamp Voltage Rise
A shortening of lamp life may be caused by an excessive rise in lamp voltage during life, causing the lamp to extinguish after burning for some time after ignition. The lamp reignites after a short cooling-off time, this process repeating itself (cycling).
A typical example of lamp voltage rise is shown in figure 10.3, where the relative increase in the lamp voltage of 400 W HPS lamps is given as a function of the operation time. Should the lamp voltage become larger than about

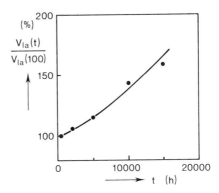

Figure 10.3 Relative increase of the average lamp voltage of standard 400 W HPS lamps as a function of burning time. (Claassens et al., 1984)

170 per cent of the one-hundred hour operation value, then the lamp extinguishes and the lamp reaches the end of its life.

Because lamp voltage rise during lamplife is such an important aspect of the HPS lamp (Jacobs *et al.*, 1978; Inouye *et al.*, 1979; Kerekes, 1979) it warrants special consideration.

The lamp voltage is determined by the electrode fall, the arc length and the electric field strength. A variation in any one of these parameters will cause the lamp voltage to change.

During operation, the surface of the electrode eventually becomes totally devoid of activator material, this having been sputtered away from the electrode during the glow phase at each start of the lamp and, subsequently, by evaporation in the stationary operation situation (Sec. 9.2.2). When, as a result of sputtering and evaporation, the first windings of the electrode contain no more activator material, the arc path will lengthen as the cathode spot shifts a few mm farther away from the electrode tip (figure 10.4). Moreover, a cathode that is totally devoid of activator material brings about a higher cathode fall than a fresh electrode, which is amply supplied with this material. However, both phenomena together – a longer arc length and a larger electrode fall – will never lead to a voltage rise of more than 10 V. Lamp voltage rise has therefore to be attributed mainly to the increase in the electric field strength.

The electric field strength in an HPS lamp with given discharge-tube diameter depends mainly on the partial vapour pressures of sodium and mercury, and therefore on the temperature and the composition of the amalgam, both of which may change during life. Physical and/or chemical binding of sodium, by reactions between the sodium and components of the sputtered or evaporated emitter material, the discharge tube material (Sec. 8.1.4) and the sealing ceramic (Sec. 8.2.2) may change the composition of the amalgam. A modified power balance at the end of the discharge tube caused by end-blackening,

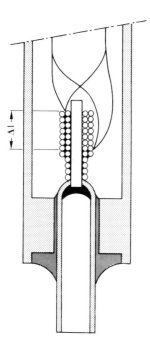

Figure 10.4 Lengthening of the arc path Δl, by a shift of the cathode spot to a place several windings away from the first winding (near the tip) during lamplife.

increased electrode losses, or a shift of the cathode spot, influences the amalgam temperature as well.

The influence of the sodium loss Δm_{Na} from the amalgam and of its rise in temperature ΔT_{am} on the partial vapour pressures of sodium and mercury can be calculated using the relations given in Sec. 5.6.

The masses of evaporated mercury and sodium in the discharge column and in the cold ends behind the electrodes are calculated from the equation of state and the temperature distribution. The plasma temperature is calculated by solving the power balance equation – the effective temperature of the vapour in the cold ends of the discharge tube is estimated to be 1800 K. The partial vapour pressures of sodium and mercury can, for a given amalgam temperature and sodium loss, be calculated by iteration from a combination of Eqs 5.12 to 5.19. From these data the field strength is derived (Sec. 5.1), while Eq. 4.26 gives the lamp voltage, for a known electrode fall V_{el}.

By way of example, the calculated lamp voltages for an experimental 50 W HPS lamp are given in figure 10.5 as functions of the sodium vapour pressure for two values of the amalgam mass and three values of the sodium loss. To relate the sodium loss to the measured values of V_{la} and $\Delta\lambda_B$ (where the self-reversal width $\Delta\lambda_B$ is a linear function of the sodium vapour pressure as described in Secs 3.2.1 and 5.5.2), the latter is given along the upper horizontal axis. For the dashed curves in figure 10.5 the amalgam temperature

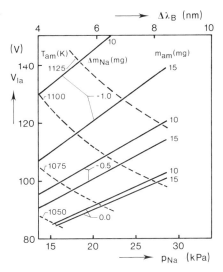

Figure 10.5 Influence of sodium loss Δm_{Na} and amalgam temperature T_{am} on the lamp voltage V_{la} of an experimental 50 W HPS lamp ($l = 27$ mm, $D = 3.3$ mm, $V_{el} = 15$ V, $x_{Na} = 0.67$). The calculated lamp voltages are given as functions of the sodium vapour pressure p_{Na} for two values of amalgam mass m_{am} (10 and 15 mg) and three values of the sodium loss Δm_{Na} (0.0, 0.5 and 1.0 mg). The temperatures on the dashed curves (1050, 1075, 1100 and 1125 K) indicate the amalgam temperatures necessary to get the sodium vapour pressure under the combined conditions of amalgam mass and sodium loss. $\Delta \lambda_B$ is the wavelength separation between the maximum in the blue wing and the centre of the sodium D-lines.

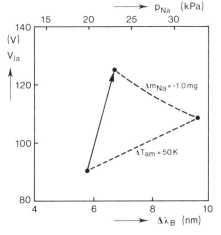

Figure 10.6 Analysis of the rise of lamp voltage V_{la} of a 50 W experimental HPS lamp after 10 000 burning hours. Successively the influences are traced of the change in the amalgam temperature ΔT_{am} and of the sodium loss Δm_{Na} on the lamp voltage V_{la} and on the sodium vapour pressure p_{Na}.
$\Delta \lambda_B$ is the measured wavelength difference between the maximum in the blue wing and the centre of the sodium D-lines. The initial amalgam mass is 15 mg in a discharge tube volume of 320 mm^3.

is constant. The rise in the lamp voltage during lamplife due to changes in amalgam temperature and sodium content in the amalgam can now be analysed with the help of figures like 10.5. To this end, only measurements of the lamp voltage V_{la} and $\Delta \lambda_B$-value are necessary. As an example, the result of such an analysis of the lamp voltage rise is given in figure 10.6. The influence is traced of the amalgam temperature increase ΔT_{am} and of the sodium loss Δm_{Na}, both with respect to the reference lamp conditions. In figure 10.6 the influences of both effects, viz. rise in amalgam temperature and sodium loss, are of about equal importance. The main cause of voltage rise for 400 W HPS lamps as found by Jacobs et al.(1978), was rise in amalgam temperature. But according to the experiments of Inouye et al. (1979), clean-up of sodium plays the major role. Which of these effects – temperature rise of the amalgam or sodium loss – is dominant, probably depends on various design parameters. Up to now it is not fully understood how the two effects depend on the design parameters.

Zollweg and Kussmaul (1983) used a similar analysis technique to find the cause of voltage rise in 400 W HPS lamps. The only difference is that they constructed graphs as presented in figure 10.5 from experiments with 400 W HPS lamps, the coldest spot temperature of which was varied using a bath of molten tin.

A number of measures can be taken to reduce the voltage rise to a minimum. One is to construct new electrodes, keeping the emitter material away from the electrode's outer surface to prevent the former from being easily sputtered away during the glow phases. Another possibility is the use of new emitter materials with lower evaporation rates. These measures (Sec. 9.2) are based on the assumption that deposits on the inner ends of the discharge tube cause a variation in the amalgam temperature and an enhanced sodium loss.

Voltage rise can also be reduced by decreasing the wall temperature to values below 1400 K, by carefully controlling the impurites in the PCA discharge tube material, or by using sapphire as the discharge tube material. These measures are all based on the reduction of sodium loss brought about by its reaction with the wall material (Sec. 8.1.4).

Another way of limiting the voltage rise (which also serves to reduce blackening of the discharge tube ends – see figure 9.8) is to increase the xenon starting gas pressure (see figure 10.7). At the same time this lowers the reignition peak after current zero (figure 5.21), so that the lamp does not extinguish until after a larger lamp voltage rise.

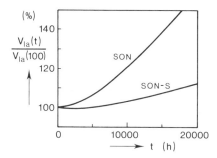

Figure 10.7 Limiting the lamp voltage rise in SON-S lamps by the increased xenon starting gas pressure. The relative increase of the average lamp voltage of 150 W HPS lamps of the standard type (SON) and of lamps with increased xenon starting gas pressure (SON-S) are given as functions of burning time.

Lumen Maintenance

The fall off in light output during lamp life is a consequence of the deterioration of the transmission of both the discharge tube and the outer bulb. The significant light loss attributable to the accumulation of dust on the light-emitting window of the luminaire will not be discussed here. The fall in light output during the early life of the lamp because of the deterioration of the transmission may be partly compensated by the increase in lamp power as a consequence of lamp voltage rise.

During the life of a lamp the transmission of the discharge tube wall is progressively reduced by blackening of the discharge tube near the electrodes and by a greying of the central part of the discharge tube (Sec. 8.1.4). Most measures outlined above for preventing excessive lamp voltage rise are also effective in maintaining the luminous efficacy of the lamp.

Light-absorbing aluminium is deposited on the inner surface of the outer bulb as a result of sublimation of the alumina (Sec. 8.1.4). This leads to a decrease in the light transmission of the outer bulb. The transmission decrease is faster for higher temperatures of the discharge-tube wall. This is illustrated in figure 10.8, where the light transmittance is given as a function of the operation time for various wall temperatures as derived by Rickmann (1977) for smooth-ground-finish alumina. As can be seen, a parameter of essential im-

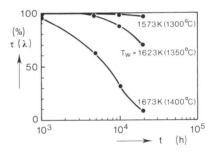

Figure 10.8 Decrease of the transmittance τ for $\lambda = 600$ nm of the aluminium layer on the outer bulb of HPS lamps as functions of the burning time t for three discharge tube wall temperatures T_w. (Rickman, 1977)

portance for lumen maintenance is the wall temperature of the discharge tube. The light transmission of the outer bulb of practical HPS lamps falls more rapidly with burning time than does that of lamps with smooth-ground-finish alumina discharge tubes. This is because surface irregularities of the discharge tube enhance the sublimation of alumina. Generally speaking, for a lifetime in excess of 10000 hours the wall temperature should not exceed about 1500 K. As a consequence, the wall loading of a discharge tube should – dependent on the thickness of the tube wall – be chosen below a certain limit. The wall temperature of an axially homogeneous discharge tube with an evacuated outer bulb is determined by the balance between the sum of the radiation absorbed by the wall and the conduction power, on the one hand, and thermal radiation on the other. If the power absorbed by the wall represents a fraction f of the electrical power dissipated in the discharge column, then the wall loading P_w as defined in Eq. (4.21), is related to the radiant exitance M by

285

$$P_w = \frac{D + 2d}{f\,D}\,M \qquad\qquad\qquad (10.1)$$

where D = (inner) diameter of the discharge tube
            d = wall thickness of the discharge tube

The radiant exitance of an alumina discharge tube depends on the temperature and thickness of the discharge tube wall (figure 8.19). From the data in this figure and Eq. 10.1 it follows that the practical HPS lamps (f ≈ 0.4, see Sec. 2.4) with tube wall thicknesses between 0.4 and 0.8 mm should not be designed with wall loadings larger than certain values depending on tube diameter and wall thickness. These values come out between 0.1 and 0.2 W mm^{-2}. In this manner the wall temperature of elongated discharge tubes is held below the 1500 K limit. However, for shorter discharge tubes larger loadings may be tolerated if a substantial part of the dissipated power is transported through the alumina from the middle of the discharge tube to the tube ends by thermal conduction. By way of example, the measured temperature distribution of the short discharge tube of a 35 W HPS lamp is given in figure 10.9. Combining Eq. (10.1) with the data from figure 8.19 one would expect a more than 100 K higher wall temperature in the middle of a long discharge tube than actually measured. For very short discharge tubes the wall temperature and not the wall loading should be chosen as design parameter.

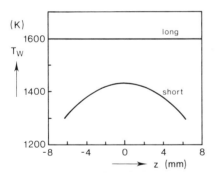

Figure 10.9 Measured wall temperature profiles of a short 35 W HPS lamp compared with the calculated temperature profile of a long discharge tube at about the same wall loading (0.2 W mm^{-2}), wall thickness (d = 0.6 mm) and diameter (D = 3.3 mm). The electrode spacing of the short discharge tube was 13 mm; Z is the distance from the middle of the discharge tube.

10.1.3 Luminous Efficacy

The luminous efficacy of an HPS lamp depends primarily on three types of lamp parameters. First there are those like sodium vapour pressure and buffer gas pressure, an increase in which causes the luminous efficacy to pass through a maximum. Second there are those like discharge tube diameter and wall loading, an increase in which causes the luminous efficacy to continuously increase (with optimal filling of the discharge tube). The effects of these two groups of parameters will be discussed separately. Finally, there are those lamp parameters that can almost be optimised independently of the ones already mentioned. For example, parameters defining the transmission of the discharge tube and the electrode losses (Chapters 8 and 9 respectively). Of these parameters, only the influence of the wall thickness on the luminous efficacy will be explained in this section.

Sodium Vapour Pressure
As already indicated in figure 1.8, there exist two maxima in the curve giving the luminous efficacy as a function of the sodium vapour pressure. The maximum at the highest value of the sodium vapour pressure describes the situation in standard HPS lamps. The sodium vapour pressure, and thus the $\Delta\lambda$-value (the wavelength separation between the maxima of the self-reversed D-lines), has to be carefully chosen so as to maximise the luminous efficacy.

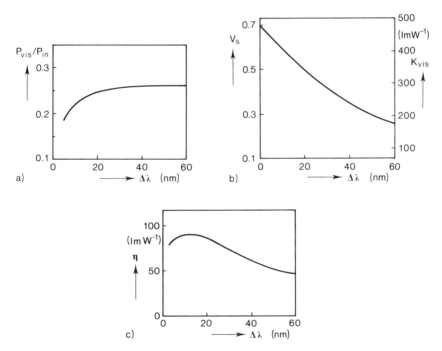

Figure 10.10 Influence of the wavelength difference $\Delta\lambda$ between the maxima of the self-reversed D-lines (the sodium vapour pressure) on
a) the efficiency for generation of visible radiation P_{vis}/P_{in},
b) the luminous efficiency V_s of this visible radiation, and
c) the luminous efficacy η.
Data are given for an experimental 150 W HPS lamp without buffer gas.

287

This is illustrated in figure 10.10 (a, b and c). For sodium vapour pressures below this optimum value the amount of visible radiation (P_{vis}) decreases, while for pressures above this point the luminous efficiency (V_s) of this visible radiation decreases. This may be explained as follows. As the sodium vapour pressure increases, starting from a low value, the radiant efficiency increases (Chapter 4, figure 4.14) and with it P_{vis} (figure 10.10a). Above the optimum sodium vapour pressure the radiant efficiency is found to increase very slowly. The broadening of the D-lines, however, increases in such a pronounced way that the spectrum becomes increasingly more badly matched to the spectral luminous efficiency curve $V(\lambda)$ of the eye, leading to a decrease in the luminous efficiency of the emitted radiation (figure 10.10b). Generally, $\Delta\lambda$ is chosen between 8 and 12 nm, for which values the luminous efficacy is near a maximum (figure 10.10c).

Kind and Pressure of the Buffer Gas

When a buffer gas (xenon and mercury, being the most appropriate ones for a high luminous efficacy as stated in Chapter 5) is introduced into the discharge tube, the optimum $\Delta\lambda$-value will differ only slightly from that existing when such a gas is absent. The fraction of the input power lost by heat conduction decreases by increasing the relative value of the buffer gas pressure. At the same time, due to the extra broadening of the sodium lines, the fraction of the radiation emitted in the sodium D-lines increases (see Sec. 5.3).

Figure 10.11a shows the efficiency with which visible radiation is generated as a function of the ratio of the pressure of the buffer gas in use to the optimum sodium vapour pressure. It is evident that both buffer gases lead to an increase in efficiency in this respect. Furthermore, there is nothing to be gained from increasing the pressure ratio to values higher than 20 to 30 to obtain a higher efficiency for those HPS lamps where either mercury or xenon is added as the buffer gas.

In addition, the influence of the buffer gas on the luminous efficiency of the visible radiation must be considered. Figure 10.11b gives the luminous efficiency of the visible radiation as a function of the ratio of the buffer gas pressure to the optimum sodium vapour pressure. This figure shows an interesting difference in the influence of the buffer gases mercury and xenon, in that the decrease in luminous efficiency with increasing buffer gas pressure is clearly more pronounced for the former than for the latter. As was mentioned in Chapter 5, this is caused by the different influences that these two buffer gases have on the spectrum.

A combination of figures 10.11a and b (figure 10.11c) indicates that the luminous efficacy reaches a broad optimum at pressure ratios $p_{Xe}/p_{Na} \approx 10-30$ and $p_{Hg}/p_{Na} \approx 10-20$. The latter ratio corresponds to a sodium mole fraction

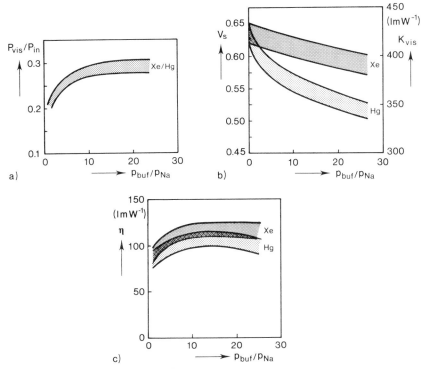

Figure 10.11 Influence of the ratio p_{buf}/p_{Na} of the buffer gas pressure of xenon and mercury respectively to the sodium vapour pressure on
a) the efficiency for generation of visible radiation P_{vis}/P_{in},
b) the luminous efficiency V_s of this visible radiation, and
c) the luminous efficacy η.
Data are given for experimental 150 W HPS lamps at a constant $\Delta\lambda$ (Glenny et al., 1978). P_{vis} is obtained by integration of the spectral power distribution between 380 and 740 nm.

in the amalgam composition of between 60 and 70 per cent, as can be deduced from figure 5.28. The same conclusions can be derived from the experimental data of Denbigh (1974), who found a broad optimum in the luminous efficacy as a function of the sodium content in the amalgam.

Using xenon instead of mercury as the buffer gas, lamps with higher luminous efficacy values can be realised. However, apart from the ignition problems associated with the use of xenon, there are the following disadvantages to be considered

– the shift in the chromaticity coordinates reveals a displacement of the colour point away from the black body locus (Sec. 5.2.2);
– the lower field strength requires longer discharge tubes, mostly with smaller (internal) diameters, so giving a lower radiant efficiency for a given lamp voltage and lamp power.

289

These disadvantages can be fully cancelled out by using a combination of both buffer gases. As was discussed in Chapter 5, the addition of mercury to a sodium-xenon discharge with a high xenon pressure increases the electric field strength and moves the colour point back to the black-body locus without seriously affecting the very high luminous efficacy. The buffer gas pressure of xenon is limited to between 20 and 40 kPa (at room temperature) by the required ignition characteristics (Chapter 6).

Wall Loading and Diameter of Discharge Tube
The conclusions that can be drawn from the data given in Chapter 4 (figure 4.15) are that
- the higher the wall loading, the more efficiently is the radiation generated until the efficiency approximates a limiting value;
- the larger the diameter of the discharge tube, the more efficiently is the radiation generated, until the efficiency approximates a nearly constant value.

The same conclusions can be drawn with respect to the luminous efficacy. This knowledge gives the designer the possibility to use the arc tube diameter and the electrode distance to satisfy the requirements with respect to the electrical characteristics (lamp voltage) and lamplife (wall temperature) without seriously affecting the optimum value for the luminous efficacy.
For HPS lamps with low power ratings, having a small diameter of the discharge tube, the luminous efficacy can be seriously decreased by this smaller tube diameter. Also, as follows from the data given in figure 10.9, the one-to-one relationship between wall loading and wall temperature is no longer valid for the short discharge tubes used in such lamps. Therefore, the design of lamps of low power rating requires more experimental work than does that of lamps of higher power rating.

Wall Thickness
The thinner the wall of the discharge tube, the higher the wall temperature and the wall transmittance and the higher the luminous efficacy.
Several authors have recently reported their findings as regards the influence of the wall temperature on the luminous efficacy. They deduced this influence from experiments on discharge tubes heated in a furnace (Waymouth and Wyner, 1981), on discharge tubes with widely-varying wall thicknesses (van Vliet and de Groot, 1982 and 1983), and on discharge tubes mounted in an outer bulb filled with gases of various pressures (Gillard and Ingold, 1983) and of different thermal conductivities (Denbigh *et al.*, 1983). The influence can also be deduced from model calculations (van Vliet and de Groot, 1982 and 1983; Gillard and Ingold, 1983; Denbigh *et al.*, 1983). In spite of the large differences between some of the experimental results, most of the au-

thors agree that the luminous efficacy increases by about 4 per cent per 100 K increase in wall temperature. As follows from the figures 4.16 and 3.11, this increase is due partly to the enhanced radiant efficiency and partly to the enhanced Na-D line radiation with respect to the non-resonance radiation, which together result in a higher luminous efficacy.

The wall thickness influences not only the temperature of the wall but also its optical properties. To achieve low wall temperatures, thick – low light transmission – walls have to be used, while to achieve high wall temperatures, thin – high light transmission – walls are called for. Thus the influence of the wall thickness is even stronger than may be expected from the roughly 4 per cent per 100 K increase in wall temperature, mentioned in the foregoing paragraph.

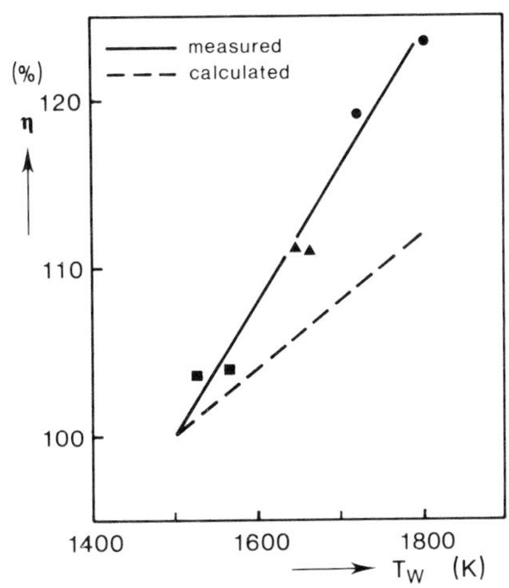

Figure 10.12 Relative increase in the luminous efficacy η of 70 W 'pure' sodium discharge lamps without buffer gas for higher wall temperatures. The calculated curve is achieved by combination of the models given in Chapters 3 and 4 (van Vliet and de Groot, 1982, 1983). Measured wall thicknesses:

● = 0.3 mm
▲ = 0.4 mm
■ = 0.6 mm

In figure 10.12 the measured relative increase in luminous efficacy of a series of experimental low power HPS lamps is given as a function of the wall temperature. The wall thicknesses of these lamps were 0.3 mm, 0.4 mm and 0.6 mm, the other parameters being kept approximately constant by adjusting the cold-spot temperatures. From this figure it can be deduced that by halving the wall thickness from 0.6 to 0.3 mm, the efficacy increases by about 20 per cent. This increase is due partly to the increase in wall temperature (12 per cent) and partly to the improved transmission of the PCA tube (8 per cent). However, for a given design of discharge tube, the lower limit of the wall thickness is given by the requirement that the wall temperature may not exceed a value of 1400 or 1500 K to ensure a reasonably long lamplife (Sec. 10.1.2). Practical values for the wall thickness of PCA discharge tubes lie between 0.6 and 0.8 mm.

10.1.4 Light Quality

It will be clear that when designing an HPS lamp for exterior lighting the quality of the light emitted is of minor importance compared to that needed for interior lighting. The light quality of a source is generally defined by its colour appearance – as expressed by its chromaticity coordinates and the consequent correlated colour temperature and by its colour rendering properties (Chapter 1). For interior lighting the quality of the light must be comparable to that given by an incandescent lamp, and this can be achieved by increasing the sodium vapour pressure (Chapters 1 and 2).

Mizuno *et al.* (1971) measured the chromaticity coordinates for various sodium and mercury vapour pressures by varying the amalgam temperature and composition. Their results are shown in figure 10.13. From this figure it ap-

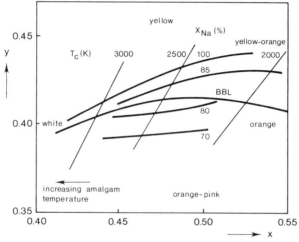

Figure 10.13 Part of the CIE chromaticity diagram showing the chromaticity coordinates x and y of HPS lamps (400–500 W) with various mole fractions x_{Na} of sodium in the amalgam.
The directions of the shift of the chromaticities as a consequence of increasing amalgam temperature is indicated. BBL = black body locus. The data in this figure were deduced from experimental data of Mizuno et al. (1971).

pears that the chromaticity coordinates are strongly influenced not only by the amalgam temperature but also by the amalgam composition. For a higher colour temperature the amalgam temperature should be increased; for a better colour appearance (viz. a colour point lying on or near the black body locus) the amalgam composition should contain a sodium mole fraction of about 80 per cent.

Improving the light quality by increasing the correlated colour temperature and the colour rendering index means that the luminous efficacy decreases. This is indicated in figure 10.14 for a 400 W HPS lamp. The decrease in

292

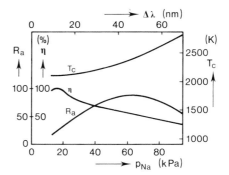

Figure 10.14 The dependence of the general colour rendering index R_a, the correlated colour temperature T_c and the relative luminous efficacy on the sodium vapour pressure p_{Na} ($\Delta\lambda$-value) of a 400 W HPS lamp. $\Delta\lambda$ is the wavelength separation between the maxima of the self-reversed D-lines. (Beijer et al., 1970, 1974)

luminous efficacy at higher sodium vapour pressures is entirely due to the decrease in the luminous efficiency of the visible radiation (figure 2.10). In general, a maximum colour rendering index of 85 is achieved at $\Delta\lambda$-values between 40 and 50 nm at the cost of about a 40 per cent loss in luminous efficacy. The correlated colour temperature has then reached a value of between 2300 and 2500 K.

Other possible ways of achieving a higher colour quality have been investigated. Lamp improvements can, theoretically at least, be brought about by adding pure metals to the sodium discharge (Hanneman, 1971; Rigden, 1972; Koedam et al., 1972), but in practice the choice available is very limited. Only elements having excitation energies near to that of sodium and capable of filling out the sodium spectrum need to be considered. Furthermore, to be suitable, the elements must have a significant vapour pressure at temperatures where the sodium vapour pressure is ca. 10 kPa. Until now, no succesfull metal additives have been reported. Metal halide lamps, in which the dosed salt mixture contains sodium iodide, fall outside the scope of this book.

An enhanced colour temperature can be achieved by raising the plasma temperature, and this can be accomplished by modifying the lamp-current waveform. This possibility which is described in more detail in Sec. 7.1.2, has not yet been employed in practice.

10.2 Examples of Lamp Design

As mentioned in the introduction to this chapter the design of HPS lamps has always to be a compromise taking into consideration the intended use of the lamp. Four groups of lamps can be identified

1) Standard lamps, in which a high luminous efficacy is combined with a long useful life.
2) Improved standard lamps, obtained by using a high xenon pressure besides mercury as buffer gas.

3) Those intended for direct replacement of HPMV lamps, in which the design guideline was to achieve the lowest possible breakdown voltage.
4) Those lamps currently under development for applications where a better light quality is required.

10.2.1 Dimensions and Filling of the Discharge Tube

The dimensions and filling of the discharge tube cannot be freely chosen. For reasons already mentioned earlier in this chapter, there are certain design parameters that have first to be chosen according to the intended application (Sec. 10.1)

- The lamp power and lamp voltage should be chosen according to the rated lumen output and the lamp circuit.
- To obtain a long lamp life the wall loading must be limited to about 0.20 W mm^{-2} for standard HPS lamps and to about 0.15 W mm^{-2} for improved-colour lamps.
- The wavelength separation between the maxima of the self-reversed D-lines should be about 10 nm for maximum efficacy and about 45 nm for the highest possible colour rendition.
- During operation the sodium content in the amalgam should be 60 to 70 mole per cent for maximum efficacy and about 75 to 85 mole per cent to get the colour point near the black body locus for colour-improved lamps.
- The starting gas pressure – or (high) xenon buffer gas pressure – should be chosen in accordance with the starting aids available for ignition.

The geometry of the rotationally symmetrical discharge tube, as determined by the electrode distance l and the arc-tube (internal) diameter D, can now be found by solving the two equations

$$P_w = \frac{P_{la} - P_{el}}{\pi D l} \tag{10.2}$$

and

$$V_{la} = \frac{0.9}{\alpha} E_{d.c.} l + V_{el} \tag{10.3}$$

The value $E_{d.c.}$ is the calculated field strength for a d.c. discharge as given in Chapter 5. The lamp power factor α_{la} is taken to be a constant. The sodium and mercury vapour pressures necessary for the calculation of the field strength are determined from the prescribed $\Delta\lambda$-value, the amalgam composition and the coldest spot temperature by solving the set of implicit equations given in Chapter 5, viz.

$$p_{Na} = p_{Na}(x_{Na}, T_{am}) \qquad (10.4)$$
$$p_{Hg} = p_{Hg}(x_{Na}, T_{am}) \qquad (10.5)$$
$$\Delta\lambda = \Delta\lambda(p_{Na}, p_{Hg}, p_{Xe}, D) \qquad (10.6)$$

The lamp current and the electrode losses satisfy the relations

$$P_{la} = \alpha_{la}\, I_{la}\, V_{la} \qquad (10.7)$$
and $\qquad P_{el} = I_{la}\, V_{el} \qquad (10.8)$

where the electrode fall V_{el} can be derived (approximately) from figure 9.10 if the electrode dimensions are known. With these equations (viz. Eqs 10.2 to 10.8) the unknown design parameters and properties, such as l, D, p_{Na}, p_{Hg}, T_{am}, I_{la} and P_{el} can be calculated for any HPS lamp if the application area is known.

By way of example, the dimensions of the discharge tube for maximum efficacy and long lamp life (viz. $\Delta\lambda = 10$ nm and $P_w = 0.18$ W mm^{-2}) have been calculated using the above equations. The calculated discharge tube diameters given in figure 10.15 are in close agreement with the experimental data

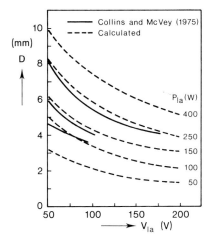

Figure 10.15 Discharge tube diameters D for standard HPS lamps calculated according to Eqs (10.4)–(10.10) as functions of the lamp voltage V_{la} for various lamp power ratings P_{la}. The sodium mole fractions in the amalgam x_{Na} = 0.74. The experimental data are from Collins and McVey (1975a).

obtained by Collins and McVey (1975a). As can be seen from this figure, an increase in lamp voltage or a decrease in lamp power calls for a reduction in tube diameter. The radiant efficiency of the discharge decreases with a reduction of the tube diameter (see Sec. 4.2.2) and this, together with the relatively higher electrode losses at lower lamp powers, are the main factors responsible for the lower luminous efficacy of lamps with low power ratings.

295

Next, the discharge tube diameter and the electrode distance for colour-improved 100 W HPS discharge lamps are calculated. The results are given in figure 10.16 a and b. The more the colour rendition is improved by choosing a higher $\Delta\lambda$-value, the larger must be the diameters of the discharge tube and the smaller the electrode distance.

Starting from Eqs (10.2) and (10.3), simple relations for the (internal) tube diameter and the electrode distance can be derived, assuming that for the field strength E the following simplified relation according to Eq. (4.32) holds

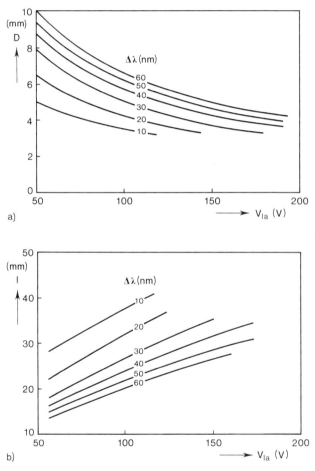

a)

b)

Figure 10.16 Values of (a) the discharge tube diameter D and (b) the electrode spacing l of HPS lamps calculated according to Eqs (10.4)–(10.8) as functions of the lamp voltage V_{la} for various $\Delta\lambda$-values (viz. the wavelength difference between the maxima of the self-reversed D-lines, which is a measure for the colour quality). The lamp power is 100 W, the wall loading 0.2 Wmm^{-2} and the sodium mole fraction in the amalgam $x_{Na} = 0.75$.

$$E = b \, D^{-a} \tag{10.9}$$

where a and b are constants.

The arc-tube diameter D and the electrode distance l are then given by

$$D = \left(\frac{0.9 \, b}{\alpha_{la} P_w} \frac{P_{la} - P_{el}}{V_{la} - V_{el}} \right)^{\frac{1}{a+1}} \tag{10.10}$$

$$l = \left(\frac{\alpha_{la}(V_{la} - V_{el})}{0.9 \, b} \right)^{-\frac{1}{a+1}} \left(\frac{P_{la} - P_{el}}{P_w} \right)^{\frac{a}{a+1}} \tag{10.11}$$

With the aid of these practical relations, the discharge tube dimensions for lamps with other power rating and voltage values can easily be deduced from two established designs. Elenbaas (1938) formulated similar relations for HPMV discharge tubes.

10.2.2 Complete HPS Lamp

The final performance of the HPS lamp is significantly influenced by the way the discharge tube is incorporated in the outer bulb (figure 10.17).

The discharge tube should be supported in the outer bulb by a frame in such a way that the passage of light is obstructed as little as possible. The tube should also be protected against shocks (for example by support springs),

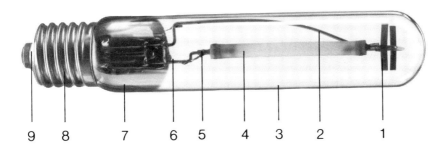

9   8   7   6   5   4   3   2   1

Figure 10.17 Example of a complete HPS lamp (SON-T)
1. *Support springs*
2. *Twisted support/lead-in wire for improved optical characteristics*
3. *Tubular or ovoid outer bulb impervious to air and other atmospheric influences*
4. *Translucent alumina discharge tube*
5. *Expansion unit to eliminate temperature stresses on welds and discharge tube*
6. *Lead-in wire/support*
7. *Getter to ensure high vacuum throughout life*
8. *E27 or E40 screw base*
9. *Porcelain insulating disc in base.*

and temperature stresses on welds eliminated by the use of expansion units. The outer bulb is generally evacuated to reduce the heat losses from the discharge tube and to eliminate corrosion of niobium parts by air. To maintain a sufficiently high vacuum throughout the life of the lamp a getter is mounted in the outer bulb. If a barium getter is used, then at the end of the production process the getter material is evaporated onto the outer bulb near the screw base with the aid of an RF coil. The barium is capable of removing the various gases liberated from the lamp parts during operation, especially the hydrogen (McGrath and Connolly, 1980). The small amounts of methane, helium and argon found in the outer bulbs of HPS lamps do not significantly influence the lamp characteristics (Meyer *et al.*, 1980)

– Methane (CH_4) is not removed by the getter but decomposes at the high operating temperature of the discharge tube, the hydrogen then being removed by the getter (Ward, 1980);

– Helium diffuses from the air through the glass of the outer bulb, when hot, and accumulates until its pressure approaches the partial pressure in the atmosphere;

– Argon is a residual of the remainder of air that was in the outer bulb prior to the flashing of the getter.

To make the HPS lamp suitable for use with specially designed optical systems, the clear hard-glass outer bulb is tubular in shape. Where simpler and less costly optical systems are involved ovoid outer bulbs are used. In this case the inside of the outer bulb is normally electrostatically coated with a uniform layer of calcium pyrophosphate powder. The use of a diffusing powder gives the outer bulb surface of the lamp a lower surface luminance so that there is less glare; because the radiation from the HPS discharge tube contains almost no ultraviolet radiation, a fluorescent powder on the inside of the outer bulb would serve no useful purpose. The ovoid, coated versions of the HPS lamp can also be used in optical systems designed for HPMV lamps.

For some indoor applications, especially in those areas where the soiling rate is high and/or cleaning is a problem (such as in industrial halls), another type of hard-glass outer bulb with a built-in reflector can be used. An internal reflector of titan oxide ensures that soiling cannot reduce its efficiency, and the shape of the outer bulb prevents accumulation of dirt on the lamp window. The advantage of this type of outer bulb is that the lamp directs the light to where it is needed without the use of an expensive optical system.

The screw base should be designed to withstand the high voltage used for the ignition of the HPS lamp. More specifically, the insulating properties of the porcelain disc in the screw base should be appropriate to the maximum permissible ignition voltage specified for the lamp.

The position of the starter, and especially that of the starting aids employed in certain types of HPS lamps, also serve to determine the final performance of the complete HPS lamp.

10.2.3 HPS Lamp Types

Standard HPS Lamps

The standard version of the HPS lamp is characterised by a high luminous efficacy combined with a long lamplife. It is normally provided with an outer bulb that is either ovoid and coated or tubular and clear.

Ignition of these standard HPS lamps is mostly performed with the help of an external electronic starter. However, some of the low-power-rating types are provided with a built-in glow-switch (starter in series with a bimetal-switch) connected across the discharge tube (colour plate 7a).

As shown in figure 10.18, the luminous efficacy of HPS lamps decreases for the lower lamp powers. This is caused by the lower radiant efficiency and the relatively larger electrode losses of lamps with lower power ratings.

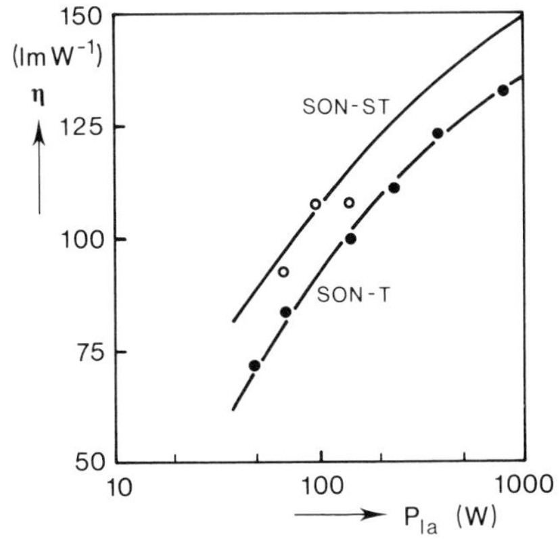

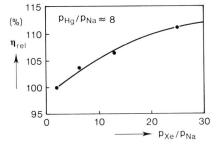

Figure 10.18 Luminous efficacy η of standard HPS lamps (SON-T) and of HPS lamps with an increased xenon pressure (SON-ST), both in tubular outer bulbs for various lamp power ratings P_{la}.

Figure 10.19 Increase in relative luminous efficacy η_{rel} by increasing the xenon pressure.
The xenon pressure p_{Xe} is given relative to the sodium vapour pressure p_{Na}.
Data apply for experimental 150 W HPS lamps where the mercury vapour pressure p_{Hg} exceeds the sodium vapour pressure by a factor of 8. (Jacobs, 1980)

HPS Lamps with High Xenon Pressure

By adding xenon gas at high pressure to the discharge of the standard version of the HPS lamp an improvement of 10–15 per cent in luminous efficacy is achieved (figures 10.18 and 19). Despite the use of starting aids, as shown in figure 6.11 and figure 9.8, the xenon pressure is limited to values of between 20 and 40 kPa (at room temperature) in order to ensure ignition on the external, electronic starter device. With such a high xenon pressure the warm-up time (time needed for the lumen output to reach 80 per cent of its nominal value) is reduced by a factor of two to between 2 and 3 minutes. But more important is the fact that the maintenance of the lumen efficacy is slightly improved (figure 10.20), and the lamplife is lengthened because a higher xenon pressure reduces the reignition voltage and the voltage rise during lamplife (figure 10.7).

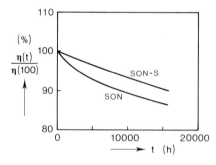

Figure 10.20 *Improved maintenance of the luminous efficacy η of the 150 W SON-S lamp relative to that of the 150 W SON lamp brought about by an increase in the xenon gas pressure.*

The improved luminous efficacy and the lengthened lamp life would make it possible to decrease the lamp power to obtain a light output equal to that of the standard lamp type, and to increase the lamp voltage, so as to reduce the ballast losses compared with those of the standard version. However, to ensure that existing installations can easily be equipped with these improved versions of the standard HPS lamp, the lamp voltage and the lamp power have not been modified with respect to the standard type. It is only in those installations that are provided from the outset with the new types, such as the 100 W version, that the full benefits from the use of high xenon pressure are obtained.

HPS Lamps as Direct Replacements for HPMV Lamps

To enable HPMV lamps to be replaced by HPS lamps the electrical characteristics of the latter have to be compatible with those of the former, viz.
– ignition must be possible at the supply voltage prevailing without the use of an external ignitor;

300

– the lamp voltage has to be increased so as to limit the ballast current to that acceptable for the HPMV ballast.

Two types of HPS lamps have been designed to satisfy these design purposes

1) In the first type, the starting gas xenon is replaced by a neon-argon Penning mixture, and an ignition wire in the form of a curl is placed round the discharge tube (figure 6.11b), the aim being to reduce the breakdown voltage to below 90 per cent of the nominal supply voltage. Data on this lamp type (SON-H) are given in table 1.3.

2) In the second type, an ignition device is mounted within the outer bulb to generate such high voltage peaks (Sec. 6.1.1) that ignition is obtained even at high xenon pressure. These high voltage peaks must not damage the HPMV ballast. The high xenon pressure lowers the reignition voltage and decreases the lamp voltage rise so that a retrofit HPS lamp compatible with the inductive HPMV ballast is realisable (Iwai et al., 1977).

These lamps should not be used for the replacement of HPMV lamps that are stabilised with the aid of capacitive circuits (Sec. 6.2).

HPS Lamps with Improved Colour Rendering

Recently, new types of HPS lamps with improved colour rendering properties were introduced on the market. Two types can be discerned

1) HPS lamps with a moderately improved colour rendering index ($R_a \approx 50$–70) and correlated colour temperature of 2150–2200 K (Bhalla et al., 1979; Otani et al., 1980, 1982; Page et al., 1980; Inada and Kaiwa, 1981).

2) HPS lamps with maximised colour rendition ($R_a > 80$), correlated colour temperatures of above 2400 K, and a colour point preferably lying on the black body locus (Akutsu et al., 1984; Claassens et al., 1984).

The improved colour rendition is obtained by an increase in the self-reversal width of the sodium D-lines $\Delta\lambda$ to about 25 and 45 nm for types 1) and 2) respectively. This higher $\Delta\lambda$ value can be realised in two ways (see Sec. 3.1.5)

a) by increasing the sodium vapour pressure,
b) by increasing the diameter of the discharge tube.

To bring the colour point nearer to the black-body locus the sodium content in the amalgam should have a mole fraction of about 80 mole per cent (figure 10.13).

The design of the discharge tube with regard to its diameter and electrode spacing is basically the same as for the standard HPS lamp. However, there are also some striking differences in lamp design to obtain an increased coldest spot temperature.

In the first place, both ends of the discharge tube may be provided with heat shields made of tantalum or niobium. This is done where the increase in the coldest spot temperature caused by a shorter distance between the electrode tip and the discharge tube end is insufficient.

Secondly, some new discharge tube constructions are in development. The monolithic PCA tube with shrunk-on ends is shown in figure 10.21. This construction reduces thermal radiation from the ends and thereby raises the coldest spot temperature without increasing the maximum wall temperature, midway between the electrodes.

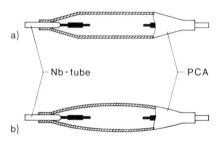

Figure 10.21 Monolithic PCA tubes with shrunk-on ends for HPS lamps with improved colour rendition. (Ogata et al., 1983)

The luminous efficacy of the HPS lamp obtained when the colour rendition is at a maximum, is about 40 per cent lower than the maximum achievable efficacy (figure 10.14). For moderate colour rendering indexes the reduction in luminous efficacy is about 10 to 15 per cent. The life of the HPS lamp having improved colour rendition is limited not only by lamp voltage rise (as in the case of the standard lamp) but also by an inadmissible deterioration in the quality of the light emitted. Sodium loss and changes in amalgam temperature during the life of the lamp can shift the chromaticity coordinates too far away from their nominal values. Therefore, in the interest of improved colour rendering, it is important to reduce sodium loss and to prevent changes of amalgam temperature from taking place. The following measures may be considered

a) The rapid reaction between sodium and the sealing ceramic at the higher amalgam temperature can be prevented by employing the modified discharge tube shown in figure 10.22. The temperature of the sealing ceramic is held below that of the coldest spot by the amalgam in the space separating the sealing ceramic from the hotter parts of the discharge tube (Claassens *et al.*, 1984).

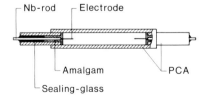

Figure 10.22 PCA tube construction for HPS lamps with improved colour rendition. The amalgam in the space between the niobium rod and the PCA separates the sealing ceramic from the hotter discharge tube parts so as to prevent a fast reaction of sodium with the sealing ceramic. (Claassens et al., 1984)

b) The influence that deposits of emitter material on the inner wall of the discharge tube around the electrodes have on the sodium loss and the amalgam temperature can be avoided by the use of electrodes not containing stored emitter material (Sec. 9.1).

c) Rapid sodium loss can be prevented by decreasing the wall temperature, preferably to below 1400 K, by employing a lower wall loading (Mizuno *et al.*, 1971) or by adopting thicker walls. The wall temperature can also be lowered by the use of a high xenon pressure, since this reduces the conduction losses.

10.3 Final Remarks

Developments involving the HPS lamp promise a number of improvements.

To begin with, the present range of HPS lamps (table 1.3) will be extended to include lower power ratings. This is necessary to satisfy the desire for the lower luminous fluxes (down to about 2 klm) needed for the lighting of residential areas and for security lighting (see figure 1.16). Because of the need for energy saving the low power HPS lamp will replace the HPMV lamp in many application areas.

At the same time, the luminous efficacy and the lifetime of the standard types will continuously be increased. This will be achieved by making use of improved discharge tube materials, such as single-crystalline alumina, sintered yttria or improved polycrystalline alumina (Chapter 8). Lamp technology will also be improved. One may also expect that the range of HPS lamps featuring a high xenon pressure will be extended to include other power ratings. In the course of time these lamps, with their higher luminous efficacy, longer lifetime and better lumen maintenance, will replace those of the standard type.

Other possible ways of improving future HPS lamps have been suggested.

Under certain conditions, the use of a gaseous atmosphere in the outer bulb makes it possible to employ higher wall temperatures in the discharge tube (Sec. 8.1.4). Higher loaded lamps with a filling gas in the outer bulb may make it possible to produce compact, high wattage HPS lamps. A gaseous atmosphere in the outer bulb may also be combined with the use of thinner-walled discharge tubes, the advantage of which has already been discussed in Sec. 10.1.3.

A large part of the input power is lost due to thermal radiation, and the use of thin reflecting layers on the inner wall of the outer bulb, as proposed by Cayless and Clarke (1963), can selectively reflect this thermal radiation back toward the discharge tube. Such reflecting layers, specially for low power ratings, could successfully be used to influence the power balance of the discharge tube (coldest spot temperature).

In the near future new types of HPS lamps with improved light quality compared to that of the standard HPS lamp will be added to the existing range. Production tolerances on such discharge lamps, their ballasts, and deviations of the mains voltage from the nominal value are more critical with respect to their performance than for standard lamps. This less satisfactory performance is due to the fact that these lamps are operated at saturated sodium and mercury pressures. Small changes in the coldest spot temperature are strongly correlated with fluctuations in the sodium vapour pressure and thus with the quality of the light emmitted. To solve these problems in future HPS lamps, especially those with highly-improved light quality, two ways are open

- Electronic regulaton of the lamp voltage. This would result in a sodium vapour pressure that is independent of production tolerances and mains voltage deviations.
- Finding the means to operate the HPS lamp with an unsaturated sodium-mercury vapour pressure (Hida *et al.*, 1979).

HPS lamps possessing a light quality approximating that of incandescent lamps will become of importance for indoor lighting purposes (colour plate 7), especially since the relatively small discharge tubes of the low-power lamps will permit the construction of spotlights having even narrower beams than those produced by the incandescent (bowl) reflector lamps. Their energy consumption and heat radiation will then be significantly lower than for incandescent lamps. This example shows clearly that future developments will be more directed toward improved integration of luminaire, ballast and lamp.

In view of these future prospects, it will be clear that the HPS lamp offers many possibilities for further development and will provide attractive solutions for many new lighting applications.

References

Abramowitz, M. and Stegun, I.A., *Handbook of mathematical functions,* Dover, New York (1965).

Akutsu, H., *The discharge plasma properties of high-pressure sodium lamps.* J. Illum. Eng. Inst. Jpn., vol. 59, pp. 498-507 (1975).

Akutsu, H. and Saito, N., *Energy balance of a high-pressure sodium arc tube.* J. Light & Vis. Env., vol. 3, no. 2, pp. 11-17 (1979).

Akutsu, H., Watarai, Y., Saito, N. and Mizuno, H., *A new high-pressure sodium lamp with high color acceptability.* J. Illum. Eng. Soc., vol. 13, pp. 341-349 (1984).

Akutsu, H., *Trends in HPS lamp technology.* Light. Res. and Technol., vol. 16, pp. 73-84 (1984).

Ali, A.W. and Griem, H.R., *Theory of resonance broadening of spectral lines by atom-atom impacts.* Phys. Rev., vol. 140A, pp. 1044-1049 (1965); vol. 144A, p. 366 (1966).

Anderson, J.M., *Temperature determination in high-pressure sodium discharges by detection of rf thermal noise.* J. Appl. Phys., vol. 46, pp. 1531-1534 (1975).

Anderson, P. and Miles, E.E., *High-pressure sodium lamps – The next generation.* National Lighting Conference, Cambridge, pp. 55-57 (1982).

Austin, B.R., *Lamp life performance of high-pressure sodium lamps in the city of London – Report No. 4.* Public Light, vol. 41, pp. 85-87 (1976).

Bartels, H., *Eine neue Methode zur Temperaturmessung an hochtemperierten Bogen-saulen* (A new method of temperature measurements in high-temperature arcs). Z. Phys., vol. 127, pp. 243-273 (1950), vol. 128, pp. 546-574 (1951).

Bartels H., *Über Kontinua und 'verbotene' Serien im Natriumbogenspektrum* (On the continuum and 'forbidden' series in the spectrum of the sodium arc). Z. Phys., vol. 73, pp. 203-215 (1932).

Beijer, L.B., Broerse, P.H., Davies, I.H. and Holmes, T., *Brighter, whiter and mightier – Light sources for public lighting.* Public Lighting, vol. 35, pp. 173-186 (1970).

Beijer, L.B., Boort, H.J.J. van and Koedam, M., *Vergelijking van lagedruk- en hoge-druk-natriumgasontladingslampen* (Comparison of low-pressure and high-pressure sodium gas discharge lamps). Electrotechniek, vol. 52, pp. 86-94 (1974a).

Beijer, L.B., Boort, H.J.J. van and Koedam, M., *The sodium discharge lamp.* Light. Design & Appl., vol. 4, pp. 15-24 (July 1974b).

Beijer, L.B., Cornelissen, G.C. and Jacobs, C.A.J., *Hoher Lichtstrom auch bei kleiner Leistung* (Higher light output also at lower power). Electrotechniek, vol. 59, pp 20-21 (1977).

Benett, S.M. and Griem, H.R., *Calculated Stark broadening parameters for isolated spectral lines from the atoms helium through calcium and cesium.* Technical Report 71-097, University of Maryland, USA (1971).

Bezemer, J., Bie, J.R. de and Heyden, R.L.A. van der, *Halogen incandescent lamp incorporating a coiled tungsten ribbon.* Light. Res. and Technol., vol. 10, pp. 167-168 (1978).

Bhalla, R.S., *HID lamp electrode comprising barium-calcium niobate or tantalate.* US Patent 4.321.503 (1978).

Bhalla, R.S., *Improved end seals for high-pressure sodium lamp arc tubes.* J. Illum. Eng. Soc., vol. 8, pp. 86-89 (1979a).

Bhalla, R.S., *Improved electron emission materials for HID lamps.* J. Illum. Eng. Soc., vol. 8, pp. 174-178 (1979b).

Bhalla, R.S., Larson, D.A. and Unglert, M.C., *HPS lamp with improved color rendering.* J. Illum. Eng. Soc., vol. 8, pp. 202-206 (1979).

Bhalla, R.S., *Materials considerations for improved color high pressure sodium lamps.* I.E.S. Conf. USA (1980).

Bhattacharya, A.K., *Measurement of breakdown potentials and Townsend ionization coefficients for the Penning mixtures of neon and xenon.* Phys. Rev. A, vol. 13, pp. 1219-1225 (1976).

Bommel, W.J.M. van and Boer, J.B. de, *Road Lighting.* Philips Technical Library/ Kluwer Technische Boeken B.V., Deventer (1981).

Bruijs, P.C.M.N. and Schellen, J.A.T., Private communication (1978).

Burggraaf, A.J. and Velzen, H.C. van, *Glasses resistant to sodium vapor at temperatures to 700 °C.* J. Am. Ceram. Soc., vol. 52, pp. 238-242 (1969).

Cahoon, H.P. and Christensen, C.J., *Effect of temperature and additives on the creep properties and recrystallization of aluminium oxide.* Technical Report 42, University of Utah, Salt Lake City, p. 76 (1955).

Campbell, J.H., *Initial characteristics of high-intensity discharge lamps on high-frequency power.* Illum. Eng., vol. 64, pp. 713-722 (1969).

Campbell, W.R., *Design of pulsed alkali vapor lamps utilizing alumina, yttria and sapphire envelopes.* J. Illum. Eng. Soc., vol. 1, pp. 281-284 (1972).

Campbell, R.J. and Kroontje, W., *Detection and measurement of the effect of hydrogen in high-pressure sodium discharge lamps.* IEE Conf. Publ. no. 143, pp. 393-396 (1976).

Campbell, R.J. and Kroontje, W., *Evaporation studies of the sintered aluminium oxide discharge tubes used in high-pressure sodium (HPS) lamps.* J. Illum. Eng. Soc., vol. 9, pp. 233-239 (1980).

Carlson, W.G., *Method of making tubular polycrystalline oxide body with tapered ends.* US Patent 3.907.949 (1975).

Carrington, C.G., Stacey, D.N. and Cooper, J., *Multipole relaxation and transfer rates in the impact approximation: application to the resonance interaction.* J. Phys. B: Atom. Molec. Phys., vol. 6, pp. 417-432 (1973).

306

Cayless, M.A. and Clarke, M.G., *Improvements in sodium vapour electric discharge lamps*. British Patent 937938 (1963).

Cayless, M.A., *Resonance radiation from high-pressure alkali-metal vapour discharges*. Proc. 7th Int. Conf. on Phenomena in Ionized Gases, Beograd (Gradevinska Knjiga), vol. I, pp. 651-654 (1965).

Cayless, M.A., *Notes on early HPS lamp developments*. Private communication (1982).

Cayless, M.A. and Marsden, A.M. (Eds), *Lamps and lighting*. Edward Arnolt Ltd, London (1983).

Chalek, C.L. and Kinsinger, R.E., *A theoretical investigation of the pulsed high-pressure sodium arc*. J. Appl. Phys., vol. 52, pp. 716-723 (1981).

Chamberlain, P.F., Nelson, E.H. and Swift, J.D., *The measurement of the radial temperature distribution in a high-pressure sodium vapour arc*. Proc. Xth Int. Conf. on Phenomena in Ionized Gases, Oxford (Donald Parsons), p. 193 (1971).

Chamberlain, P.F.W. and Swift, J.D., *The temperature distribution of the high-pressure sodium vapour discharge*. IEE Conf. Publ. No. 90, pp. 113-114 (1972).

Chase, M.W., Curnutt, J.L., Hu, A.T., Prophet, H., Syvernd, A.N. and Walker, L.C., *JANAF Thermochemical Tables*, Supplement 1974. J. Phys. Chem. Ref. Data, vol. 3, no. 2 (1974).

Cheng, D.Y., *Dynamics of arc ignition and cathode spot movement of thermionically emitting cathode surfaces*. J. Appl. Phys., vol. 41, pp. 3626-3633 (1970).

Chu, G.P.K., *Properties and evaluation of HPS lamp seals: materials and structures*. J. Illum. Eng. Soc., vol. 8, pp. 250-256 (1979).

CIE Publication 13.2., *Method of measuring and specifying colour rendering properties of light sources*, 2nd edition (1974).

Claassens, J.M.M., Peeters, J.I.C. and Plas, R.J.Q. van den, *Neue Entwicklungen auf dem Gebiet der Natriumdampf-Hochdrucklampen (New developments in the field of high-pressure sodium lamps)*, Tagungsberichte, Licht, Band II, paper 28 (1984).

Coble, R.L., *Transparent alumina and method of preparation*. US Patent 3.026.210 (1962).

Coble, R.L. and Burke, J.E., *Sintering in crystalline solids*. Proc. 4th Int. Symp. on the Reactivity of Solids. (J.H. de Boer *et al.* eds), pp. 38-51. Elsevier, Amsterdam (1961).

Cohen, S., Gungle, W.C., Gutta, J.J., Olsen, A.W. and Richardson, D.A., *Heat starting a high-pressure sodium lamp*. J. Illum. Eng. Soc., vol. 3, pp. 330-335 (1974).

Cohen, S. and Richardson, D.A., *The application of Penning mixture starting gases to high-pressure sodium lamps*. Light. Design & Appl., vol. 5, pp. 12-17 (September 1975).

Collins, B.R. and McVey, C.I., *Low wattage high pressure sodium vapor lamps*. US Patent 3.906.272 (1975a).

Collins, B.R. and McVey, C.I., *HPS lamps for use on HPM ballasts*. Light. Design & Appl., vol. 5, pp. 18-24 (September 1975b).

Collins, B.R. and Wenner, R.E., *Starting pulse tester for high pressure sodium systems.* J. Illum. Eng. Soc., vol. 5, pp. 195-200 (1976).

Collins, B.R. and Mule, S.A., *HPS lamp design – a diversication in design wattage and optical control.* Light. Design & Appl., pp. 26-31 (March 1977).

Compton, A.H. and Van Voorhis, C.C., *The luminous efficiency of gases excited by electric discharge.* Phys. Rev., vol. 21, p. 210 (1923).

Compton, A.H., *Glass and article made therefrom.* US Patent 1.570.876 (1926).

Compton, A.H., *Electric lamp.* US Patent 1.830.312 (1931).

Dakin, J.T. and Rautenberg, T.H., *Frequency dependence of the pulsed high-pressure sodium arc spectrum.* J. Appl. Phys., vol. 56, pp. 118-124 (1984).

Datta, R.K. and Grossman, L.N., *Calcia enhanced β-Al_2O_3 formation in high-pressure sodium discharge lamps.* High Temperature Lamp Chemistry (Zubler ed.). The Electrochemical Society, Proc. vol. 85-2, pp. 271-290 (1985).

Davies, I.F., *High-pressure sodium lamps and the quadrilateral diagram.* CIBS National Lighting Conference, Canterbury, pp. CC1-6 (1980).

Denbigh, P.L., *Effect of sodium/mercury ratio and amalgam temperature on the efficacy of 400 W high-pressure sodium lamps.* Light. Res. and Technol., vol. 6, pp. 62-68 (1974).

Denbigh, P.L. and Wharmby, D.O., *Electrode fall and electric field measurements in high-pressure sodium discharges.* Light. Res. and Technol., vol. 8, pp. 141-145 (1976).

Denbigh, P.L., *Experimental approach to high-pressure sodium lamp design.* Light. Res. and Technol., vol. 10, pp. 28-32 (1978).

Denbigh, P.L., *Extending the life of low-power high-pressure sodium lamps.* Light. Res. and Technol., vol. 15, pp. 171-178 (1983).

Denbigh, P.L., Jones, B.F. and Mottram, D.A.J., *Relationship between efficacy, arc tube temperature and power dissipation in a high-pressure sodium lamp.* J. Phys. D.: Appl. Phys., vol. 16, pp. 2167-2180 (1983).

Denbigh, P.L. and Jones, B.F., *A simple computer program for designing standard high-pressure sodium lamps.* Light. Res. and Technol., vol. 16, pp. 193-199 (1984).

Denbigh, P.L. and Molesdale, P.D., *Recent advances in the design of high pressure sodium lamps.* National Lighting Conference, Cambridge, pp. 201-207 (1984).

Denbigh, P.L., Jones, B.F. and Mottram, D.A.J., *Variation of efficacy with arc tube bore and current in a high-pressure sodium lamp.* I.E.E. Proc. Pt. A, vol. 132, pp. 99-103 (1985).

Denneman, J.W., *Low-pressure sodium discharge lamps.* I.E.E. Proc. Pt.A., vol. 128, pp. 397-414 (1981).

Dimitrijević, M.S. and Sahal-Bréchot, *Stark broadening of neutral sodium lines.* J. Quant. Spectros. Radiat. Transfer, vol. 34, pp. 149-161 (1985).

Dorgelo, E.G., *Alternating-current circuits for discharge lamps.* Philips Tech. Rev., vol. 2, pp. 103-109 (1937).

Dorleijn, J.W.F. and Heijden, R.L.A. van der, *Properties of high-pressure sodium arcs at frequencies above 50 Hz*. 33th Gaseous Electronics Conf., Oklahoma, paper LA-2 (1980a).

Dorleijn, J.W.F. and Heijden, R.L.A. van der, Private communication (1980b).

Dushman, S., *Scientific foundations of vacuum technique*. John Wiley & Sons, Inc., New York/London (1962).

Dyke, W.P. and Dolan, W.W., *Field Emission, Advances in Electronics and Electron Physics*, vol. VIII, (L. Marton ed.), Academic Press. Inc., New York (1956).

Elenbaas, W., *Electric Gaseous Discharge Device*, U.S. Patent 2.135.702 (1938).

Elenbaas W., *The high-pressure mercury vapour discharge*. North Holland (1951).

Elenbaas, W., *High pressure mercury vapour lamps and their applications*. Philips Technical Library, Eindhoven (1965).

Engel, A. von and Steenbeck, M., *Elektrische Gasentladungen* (Electrical Gas Discharges). Springer, Berlin (1934).

Fischer, D., *Decorative floodlighting*. International Lighting Review, vol. 30, pp. 103-110 (1979).

Foley, H.M., *The pressure broadening of spectral lines*. Phys. Rev., vol. 69, pp. 616-628 (1946).

Fomenko, V.S., *Handbook of Thermionic Properties*, (Samsonov Ed.), Plenum Press Data Division, New York (1966).

Forman, R., *Measurement of high-temperature thermal conductivity of Lucalox (Al_2O_3) using a heat pipe technique*. J. Appl. Phys., vol. 44, pp. 66-71 (1973).

Forman, R., *A proposed physical model for the impregnated tungsten cathode based on Auger surface studies of the Ba-O-W system*. Applications of Surface Science, vol. 2, pp. 258-274 (1979).

Frost, L.S. and Phelps, A.V., *Momentum-transfer cross sections for slow electrons in He, Ar, Kr and Xe from transport coefficients*. Phys. Rev., vol. 136, pp. 1538-1545 (1964).

Frouws, S.M., *Paschen curves in neon-argon mixtures*. Proc. 13th Int. Conf. on Phenomena in Ionized Gases, Venice, Italy, pp. 341-352 (1957).

Furuta, M., Maeno, Y. and Kobayashi, K., *A method for forming a green body for a ceramic arc tube used for a metal vapor discharge lamp and a molding die for forming said green body*. European Patent Application 0072190 (1983).

Gallagher, A., *The spectra of colliding atoms*. Atomic Physics 4 (G. zu Putlitz, E.W. Weber and A. Winnacker eds), pp. 559-574, Plenum Press, New York and London (1975).

Gallagher, A., *Metal Vapor Excimers*. Ch 5 in: *Topics in applied physics* (C.K. Rhodes ed.), vol. 30, pp. 135-174, Springer, New York (1979).

Gardon, R., *The emissivity of transparent materials*. J. Am. Ceram. Soc., vol. 39, pp. 278-287 (1956).

General Electric, *Look into the future*. Light, vol. 30 (1961).

Gillard, R.P. and Ingold, J.H., *The effect of wall temperature on sodium D-line reabsorption in high-pressure sodium arcs*. High Temperature Science, vol. 16, pp. 399-409 (1983).

Gitzen, W.H., *Alumina as a ceramic material*. The American Ceramic Society (1970).

Glenny, E.T., Jacobs, C.A.J. and Vliet, J.A.J.M. van, *New Developments in high-pressure sodium discharges*. Electr. Rev., vol. 202, pp. 30-31 (1978).

Goldsmith, A., Waterman, T.E. and Hirschhorn, H.J., *Handbook of thermophysical properties of solid materials*, vol. 3: *Ceramics*. The MacMillan Company, New York (1961).

Gould, L. and Brown, S.C., *Microwave determination of the probability of collisions of electrons in helium*. Phys. Rev., vol. 95, pp. 897-903 (1954).

Griem, H.R., *Plasma Spectroscopy*. McGraw-Hill Book Company, New York (1964).

Groot, J.J. de, *Comparison between the calculated and the measured radiance at the centre of the D-lines in a high pressure sodium vapour discharge*. IEE Conf. Publ. no. 90, pp. 124-126 (1972a).

Groot, J.J. de, *Phasenabhängige Emissions- und Absorptionsspektra einer 50 Hz-Hochdrucknatriumdampf-entladung* (Phase dependent emission and absorption spectra of a 50-Hz high-pressure sodium discharge) (Frühjahrstagung, Kiel). Verhandlungen der Deutschen Physikalischen Gesellschaft, vol. 3, p. 113 (1972b).

Groot, J.J. de, *Measurement of the temperature distribution and calculation of the total spectrum of a high pressure sodium vapour discharge* (26th Gaseous Electronics Conference, Madison). Bull. American Physical Society, vol. 19, p. 164 (1974a).

Groot, J.J. de, *Investigation of the high-pressure sodium and mercury/tin iodide arc*. Ph. D. thesis, Technical University of Eindhoven (1974b).

Groot, J.J. de and Rooijen, J. van, *The photon absorption cross-section of Na$_2$ molecules and the influence of these molecules on the spectrum of the high pressure sodium arc*. Proc. 12th Int. Conf. on Phenomena in Ionized Gases, Eindhoven (North Holland), Part I, p. 135 (1975).

Groot, J.J. de and Vliet, J.A.J.M. van, *The measurement and calculation of the temperature distribution and the spectrum of high-pressure sodium arcs*. J. Phys. D: Appl. Phys., vol. 8, pp. 651-662 (1975).

Groot, J.J. de, Vliet, J.A.J.M. van and Waszink, J.H., *The high-pressure sodium lamp*. Philips Tech. Rev., vol. 35, pp. 334-342 (1975a).

Groot, J.J. de, Janssen, M.M.H., Opstelten, J.J. and Schayk, F.H.C. van, *Measurement of spectral power distribution via a computer controlled spectroradiometer*. Proc. 18th C.I.E. Session, London, Publication CIE no 36, pp. 145-154 (1975b).

Groot, J.J. de and Vliet, J.A.J.M. van, *Determination of the Na and Hg vapour pressure in high pressure Na-Hg discharges*. (30th Gaseous Electronics Conf., Palo Alto). Bull. American Physical Society, vol. 23, p. 133 (1978).

Groot, J.J. de and Vliet, J.A.J.M. van, *Determination of the sodium and mercury vapour pressure in high-pressure sodium lamps*. 2nd Int. Symp. on Incoherent Light Sources, Enschede, Netherlands, summaries pp. 30-31 (1979).

Groot, J.J. de, Woerdman, J.P. and Kieviet, M.F.M. de, *The influence of NaNa, NaHg and NaXe molecules on the spectrum of the high-pressure sodium lamp*. To be presented at 4th Int. Symp. on the Science and Technology of Light Sources, Karlsruhe, Germany (1986).

Hanneman, R.E., Private communication to R. Hultgren (see Hultgren *et al.*, 1973) 1968.

Hanneman, R.E., Jorgensen, P.J. and Speros, D.M., *Alumina-ceramic sodium vapor lamp*. US Patent 3.453.477 (1969a).

Hanneman, R.E., Louden, W.C. and McVey, C.I., *Thermodynamic and experimental studies of the high pressure sodium lamp*. Illum. Eng., vol. 1, pp. 162-166 (1969b).

Hanneman, R.E., *Metallic vapour arc-lamp*. British Patent 1248157 (1971).

Hassan, A. and Bauer, A., *Investigation of resonance radiation of the sodium high-pressure lamp*. IEE Conf. Publ. no. 118, pp. 42-44 (1974).

Hedges, R.E.M., Drummond, D.L. and Gallagher, A., *Extreme-wing line broadening and Cs-inert-gas potentials*. Phys. Rev. A, vol. 6, pp. 1519-1544 (1972).

Hermann, G. and Wagener, S., *Die Oxyd Kathode* (The oxide cathode), Teil 1, Johann Ambrosius Barth, Leipzig (1948).

Herzberg, G., *Molecular spectra and molecular structure* I. *Spectra of diatomic molecules*. D. van Nostrand Company, Inc., New York, London (1950).

Hida, Y., Fulukubo, H., Takeji, Y. and Takatsuka, K., *High pressure sodium lamp of unsaturated vapor pressure type*. G.S. News, vol. 38, no. 2, p. 92 (1979).

Hing, P., *Interaction of alkali metal and halide vapors with ceramic materials*. J. Illum. Eng. Soc., vol. 10, pp. 194-203 (1981).

Hirayama, C. and Bhalla, R.S., *The dissociative vaporization of alkaline earth-containing mixed oxide thermionic emission materials*. J. Illum. Eng. Soc., vol. 9, pp. 240-246 (1980).

Hirayama, C., Andrew, K.F. and Kleinosky, R.L., *Activities and thermodynamic properties of sodium amalgams at 500-700°C*. Thermochimica Acta, vol. 45, pp. 23-37 (1981a).

Hirayama, C., Singleton, J.H. and Wolfe, A.L., *Emission spectra, temperature, and voltage measurements of HPS and HPM lamps at start-up*. J. Illum. Eng. Soc., vol. 10, pp. 90-95 (1981b).

Hirayama, C., Andrew, K.F. and Kleinosky, R.L., *The vapor pressures of sodium and mercury over sodium amalgams at HPS lamp operating temperatures*. J. Illum. Eng. Soc., vol. 12, pp. 66-69 (1983).

Hirschfelder, J.O., Curtis, C.F. and Bird, R.B., *Molecular theory of gases and liquids*. Wiley (1954).

Hochstim, A.R. and Massel, G.A., *Kinetic processes in gases and plasmas*. Academic Press, New York and London (1969).

Hodge, J.D., *Alkaline earth effects on the reaction of sodium with aluminium oxide*. High Temperature Lamp Chemistry (Zubler ed.). The Electrochemical Society, Proc. vol. 85-2, pp. 261-270 (1985).

Hoek, W.J. van den and Visser, J.A., *Opto-galvanic spectroscopy and thermal relaxation in high-pressure mercury and sodium arc discharges.* J. Appl. Phys., vol. 51, pp. 5292-5294 (1980).

Hoek, W.J. van den and Visser, J.A., *Diagnostics of high-pressure arc plasmas from laser-induced fluorescence.* J. Phys. D: Appl. Phys., vol 14, pp. 1613-1628 (1981).

Hogervorst, W., *Transport and equilibrium properties of simple gases and forces between like and unlike atoms.* Physica, vol. 51, pp. 77-89 (1971).

Holstein, T., *Imprisonment of resonance radiation in gases.* Phys. Rev., vol. 72, pp. 1212-1233 (1947).

Houston, W.V., *Resonance broadening of spectral lines.* Phys. Rev., vol. 54, pp. 884-888 (1938).

Hoyaux, M.F., *Arc Physics.* Springer, Berlin/New York (1968).

Huennekens, J. and Gallagher, A., *Self-broadening of the sodium resonance lines and excitation transfer between the $3P_{3/2}$ and $3P_{1/2}$ levels.* Phys. Rev. A, vol. 27, pp. 1851-1864 (1983).

Huennekens, J., Schaefer, S., Ligare, M. and Happer, W., *Observation of the lowest triplet transitions $^3\Sigma^+_g - ^3\Sigma^+_u$ in Na_2 and K_2.* J. Chem. Phys., vol. 80, pp. 4794-4799 (1984).

Hultgren, R., Orr, R.L., Anderson, P.D. and Kelley, K.K., *Selected values of Thermodynamic Properties of Metals and Alloys.* John Wiley & Sons, Inc. New York/London (1963).

Hultgren, R., Desai, P.D., Hawkins, D.T., Gleiser, M. and Kelley, K.K., *Selected values of the thermodynamic properties of binary alloys.* American Society for Metals (1973).

Hüwel, L., Maier, J. and Pauly, H., *Excited-state potentials in the Na-Hg system: Analysis of rainbow-scattering and polarization effects.* J. Chem. Phys., vol. 76, pp. 4961-4971 (1982).

IEC publication 662, *High-pressure sodium vapour lamps,* 1st edition (1979).

ILR, *International Lighting Review,* vol. 32, pp. 100-101 (1981).

Inada, A. and Kaiwa, S., *High pressure sodium lamp with improved color rendition.* J. Light & Vis. Env., vol. 5, no. 2, pp. 7-10 (1981).

Ingard, U., *Acoustic wave generation and amplification in a plasma.* Phys. Rev., vol. 145, pp. 41-46 (1966).

Ingard, U. and Schulz, M., *Acoustic wave mode in a weakly ionized gas.* Phys. Rev., vol. 158, pp. 106-112 (1967).

Inglis, D.R. and Teller, E., *Ionic depression of series limits in one-electron spectra.* Astrophys. J., vol. 90, pp. 439-448 (1939).

Ingold, J.H. and Taylor, W.L., *Correlation of high pressure sodium lamp performance with arc tube transmission properties.* J. Illum. Eng. Soc., vol. 11, pp. 223-230 (1982).

Inouye, A., Higashi, T., Ishigami, T., Nagano, S. and Shimojima, H., *The cause of lamp voltage increase of high pressure sodium lamp.* J. Light & Vis. Env., vol. 3, no. 1, pp. 1-5 (1979).

Ishigami, T., *Thermodynamic consideration on interactions between fillings and envelopes in HID lamps*. High Temperature Lamp Chemistry (Zubler ed.). The Electrochemical Society, Proc. vol. 85-2, pp. 145-155 (1985).

Iwai, I., Ochi, M. and Masui, M., *A newly designed high pressure sodium lamp*. J. Light & Vis. Env., vol. 1, no. 1, pp. 7-12 (1977).

Jack, A.G., *High pressure gas discharges as intense non-coherent light sources*. Proc. 10th Int. Conf. on Phenomena in Ionized Gases, Oxford (Donald Parsons), Invited Papers, pp. 205-229 (1971).

Jack, A.G. and Koedam, M., *Energy balances for some high pressure gas discharge lamps*. J. Illum. Eng. Soc., vol. 3, pp. 323-329 (1974).

Jacobs, C.A.J., Private communication (1978, 1980).

Jacobs, C.A.J., Sprengers, L. and Vaan, R.L.C. de, *Arc voltage control in low and high pressure sodium lamps*. J. Illum. Eng. Soc., vol. 7, pp. 125-131 (1978).

Jacobs, C.A.J. and Vliet, J.A.J.M. van, *A new generation of high-pressure sodium lamps*. Philips Tech. Rev., vol. 39, pp. 211-215 (1980).

Jeung, G., *Theoretical study on low-lying electronic states of Na_2*. J. Phys. B: At. Mol. Phys., vol. 16, pp. 4289-4297 (1983).

Johnson, P.D. and Rautenberg, T.H., *Spectral change mechanism in the pulsed high-pressure sodium arc*. J. Appl. Phys., vol. 50, pp. 3207-3211 (1979).

Jones, B.F. and Mucklejohn, S.A., *Temperature variation of sodium and mercury partial pressures over sodium amalgams*. Light. Res. and Technol., vol. 16, pp. 137-139 (1984).

Jongerius, M.J., Ras, A.J.M.J. and Vrehen, Q.H.F., *Optogalvanic detection of acoustic resonances in a high-pressure sodium discharge*. J. Appl. Phys., vol. 55, pp. 2685-2692 (1984).

Jongerius, M.J. and Ras, A.J.M.J., Private communication (1985).

Kerekes, B., *Dritte Generation der Natriumdampflampen* (Third generation of high-pressure sodium lamps). Tungsram Techn. Mitteilungen, vol. 37, pp. 1539-1546 (1978).

Kerekes, B., *Ersatz der Quecksilberdampflampe durch Natriumdampf-Hochdruck-lampe* (Replacement of mercury lamps by high-pressure sodium lamps). Tungsram Techn. Mitteilungen, vol. 42, pp. 1717-1721 (1979).

Kingery, W.D., Bowen, H.K. and Uhlmann, D.R., *Introduction to ceramics*. John Wiley & Sons, New York (1975).

Kobayashi, K. and Kaneno M., *Translucent alumina containing magnesia yttria and lanthium oxide*. US Patent 3.905.845 (1975).

Koedam, M., Opstelten, J.J. and Radielović, D., *The application of simulated spectral power distributions in lamp development*. J. Illum. Eng. Soc., vol. 1, pp. 285-289 (1972).

Kok, J. de, *High-pressure gas discharge lamp and electron emissive electrode structure therefore*. Netherlands Patent 75.073.56 (1975).

Konowalow, D.D., Rosenkrantz, M.E. and Olson, M.L., *The molecular electronic structure of the lowest $^1\Sigma_g^+$, $^3\Sigma_u^+$, $^3\Sigma_g^+$, $^1\Pi_u$, $^1\Pi_g$, $^3\Pi_u$, and $^3\Pi_g$ states of Na$_2$.* J. Chem. Phys., vol. 72, pp. 2612-2615 (1980).

Konowalow, D.D., Rosenkrantz, M.E. and Hochhauser, D.S., *Electronic transition dipole moment functions and difference potentials for transitions among low-lying states of Li$_2$ and Na$_2$.* J. Molecular Spectroscopy, vol. 99, pp. 321-338 (1983).

Kusch, P. and Hessel, M.M., *An analysis of the $B^1\Pi_u - X^1\Sigma_g^+$ band system of Na$_2$.* J. Chem. Phys., vol. 68, pp. 2591-2606 (1978).

Labelle, H.E. and Mlavsky, A.I., *Growth of controlled profile crystals from the melt: Part I – sapphire filaments.* Mat. Res. Bull., vol. 6, pp. 571-579 (1971).

Lange, H. and Skibbe, R., *Herstellung, Eigenschaften und Verwendung von transparentem Sinterkorund.* (Manufacturing, properties and application of translucent alumina). Technisch-wissenschaftliche Abhandlungen der Osram-Gesellschaft, vol. 9, pp. 203-212 (1966).

Langmuir, I., *Der Dampfdruck metallischen Wolframs.* (The vapour pressure of metallic tungsten). Physik. Zeitschr. XIV, pp. 1273-1280 (1913).

Laporte, P. and Damany, H., *High density self-broadening of the 253.65 nm mercury resonance line.* J. Quant. Spectrosc. Radiat. Transfer, vol. 22, pp. 447-466 (1979).

Lapp, M. and Harris, L.P., *Absorption cross sections of alkali-vapor molecules: I Cs$_2$ in the visible II K$_2$ in the red.* J. Quant. Spectrosc. Radiat. Transfer, vol. 6, pp. 169-179 (1966).

Laskowski, B.C., Langhoff, S.R. and Stallcop, J.R., *Theoretical calculation of low-lying states of NaAr and NaXe.* J. Chem. Phys., vol. 75, pp. 815-827 (1981).

Lee, H.E. and Cram, L.E., *On the theory of radiation-dominated wall-stabilised arcs.* J. Phys. D: Appl. Phys., vol. 18, pp. 1561-1573 (1985).

Levitskii, V.A., Hekimov, Yu. and Gerassimov, Ja.I., *Thermodynamics of double oxides, II. Galvanic-cell study of barium tungstates.* J. Chem. Thermodynamics, vol. 11, pp. 1075-1087 (1979).

Li, L. and Field, R.W., *Direct observation of high-lying $^3\Pi_g$ states of the Na$_2$ molecule by optical-optical double resonance.* J.Phys. Chem., vol. 87, pp. 3020-3022 (1983).

Light & Lighting, *Recent developments in discharge lamps.* Light & Lighting, vol. 59, pp. 166-169 (1966).

Lin, F.L. and Knochel, W.J., *Contributions to the design of the high pressure sodium lamp.* J. Illum. Eng. Soc., vol. 3, pp. 303-309 (1974).

Liu, C.S., *Heat conservation system for arc lamps.* J. Illum. Eng. Soc., vol. 8, pp. 220-225 (1979).

Lochte-Holtgreven, W., (Ed.), *Plasma diagnostics.* North-Holland (1968).

Louden, W.C. and Schmidt, K., *Intermediate pressure wall stabilized gas lamp.* US Patent 3.054.992 (1962).

Louden, W.C. and Schmidt, K., *High-pressure sodium discharge arc lamps.* Illum. Eng., vol. 55, pp. 696-702 (1965).

Lowke, J.J., *A relaxation method of calculating arc temperature profiles applied to discharges in sodium vapor.* J. Quant. Spectrosc. Radiat. Transfer, vol. 9, pp. 839-854 (1969).

Lowke, J.J., *Characteristics of radiation-dominated electric arcs.* J. Appl. Phys., vol. 41, pp. 2588-2600 (1970).

Lowke, J.J., *Predictions of arc temperature profiles using approximate emission coefficients for radiation losses.* J. Quant. Spectrosc. Radiat. Transfer, vol. 14, pp. 111-122 (1974).

Lowke, J.J., Zollweg, R.J. and Liebermann, R.W., *Theoretical description of ac arcs in mercury and argon.* J. Appl. Phys., vol. 46, pp. 650-660 (1975).

Loytty, E., *A new arc tube for HPS lamps.* Light. Design & Appl., vol. 6, pp. 14-17 (February 1976).

Luthra, K.L., *A thermochemical analysis of sodium loss reactions in high pressure sodium lamps.* High Temperature Lamp Chemistry (Zubler ed.). The Electrochemical Society, Proc. vol. 85-2, pp. 156-171 (1985)

Lyyra, M. and Bunker, P.R., *The potential energy curve of the $B^1\Pi_u$ state of Na_2.* Chem. Phys. Lett., vol. 61, pp. 67-68 (1979).

MacAdam, D.L., *Maximum attainable luminous efficiency of various chromaticities.* J. Opt. Soc. Am., vol. 40, p. 120 (1950).

Mackeown, S.S., *The cathode drop in an electric arc.* Phys. Rev., vol. 34, pp. 611-614 (1926).

Malitson, I.H., *Refraction and dispersion of synthetic sapphire.* J. Opt. Soc. Am., vol. 52., pp. 1377-1379 (1962).

Mason, E.A. and Saxena, S.C., *Approximate formula for the thermal conductivity of gas mixtures.* The Physics of Fluids, vol. 1, pp. 361-369 (1958).

Matheson, R.R., *G.E's lamp research leads to unique ceramic material.* Ceramic Age, pp. 54-58 (June 1963).

McCarthy, D.E., *The reflection and transmission of infrared materials:* III, *Spectra from 2 μ to 50 μ.* Appl. Opt., vol. 4, pp. 317-320 (1965).

McCutchen, C.W., *Drift velocity of electrons in mercury vapor and mercury vapor – CO_2 mixtures.* Phys. Rev., vol. 112, pp. 1848-1851 (1958).

McGowan, T.K., *High-pressure sodium – ten years later.* Light. Design & Appl., vol. 6, pp. 45-51 (January 1976).

McGrath, G.H. and Connolly, M., *The temperature dependence of the gettering of barium flash getters in the outer jacket of HPS lamps.* Illum. Eng. Soc. Conference (1980).

McVey, C.I., *High-pressure sodium lamp seals and recent improvements.* J. Illum. Eng. Soc., vol. 8, pp. 72-77 (1979).

McVey, C.I., *High-pressure sodium lamp technology.* IEE Proc. Pt A, vol. 127, no. 3, pp. 158-164 (1980).

McVey, C.I. and Paugh, R.L., *The unique features of high pressure sodium lamps and their systems interactions*. Light. Design & Appl., vol. 12, pp. 33-40 (June 1982).

Meyer, V.D., Wyner, E.F. and Fortucci, P.L., *The effect of outer jacket gases on the performance of HPS lamps*, J. Illum. Eng. Soc., vol. 9, pp. 175-180 (1980).

Miedema, A.R. and Dorleijn, J.W.F., *Quantitative predictions of the heat of adsorption of metals on metallic substrates*. Surface Science, vol. 95, pp. 447-464 (1980).

Miles, E.E., *High-pressure sodium lamps*. Light & Lighting, vol. 62, pp. 84-89 (1969).

Mizuno, H., Akutsu, H. and Watarai, Y., *New high-pressure sodium lamp with higher colour rendition*. CIE XVII Session, Barcelone, vol. 21B, P 71.14 (1971).

Molesdale, P., *Low wattage SON lamps*. IPLE Lighting Journal, vol. 49, pp. 168-171 (1984).

Moores, D.L. and Norcross, D.W., *The scattering of electrons by sodium atoms*. J. Phys. B: Atom. Molec. Phys., vol. 5, pp. 1482-1505 (1972).

Morse, P.M., *Vibration and sound*, 2nd ed., McGraw-Hill, New York (1948).

Moskvin, Y.V., *Photoionization of atoms and recombination of ions in the vapors of alkali metals*. Opt. Spectrosc., vol. 15, pp. 316-318 (1963).

Movre, M. and Pichler, G., *Resonance interaction and self-broadening of alkali resonance lines I. Adiabatic potential curves*. J. Phys. B: Atom. Molec. Phys., vol. 10, pp. 2631-2638 (1977).

Movre, M. and Pichler, G., *Resonance interaction and self-broadening of alkali resonance lines II. Quasi-static wing profiles*. J. Phys. B: Atom. Molec. Phys., vol. 13, pp. 697-707 (1980).

Mulder, P.P. and Koevoets, M.A.C., Private communication (1979).

Muta, A. and Tsukuda, *High-pressure sodium vapour discharge lamps*. US Patent 3.912.959 (1973).

Nassar, E., *Fundamentals of gaseous ionization and plasma electronics*. Wiley-Interscience (1971).

Nelson, E.H., *New developments in high-pressure discharge lamps*. Proc. IEE, vol. 113, pp. 668-676 (1966).

Netten, A., Private communication (1975).

Netten, A. and Kaldenhoven, B., Private communication (1976).

Neve, G. de, *High-pressure sodium lamps to replace MBF lamps in existing installations*. Light. Res. & Technol., vol. 8, pp. 157-161 (1976).

Nguyen Dat, N. and Bensoussan, M., *Starterless high pressure sodium vapour lamp*. Light. Res. & Technol., vol. 9, pp. 112-114 (1977).

Niemax, K. and Pichler, G., *New aspects in the self-broadening of alkali resonance lines*. J. Phys. B: Atom. Molec. Phys., vol. 8, pp. 179-184 (1975).

Norcross, D.W., *Low energy elastic scattering of electrons by Li and Na*. J. Phys. B: Atom. Molec. Phys., vol. 4, pp. 1458-1475 (1971).

Ogata, Y., Ikeda, T. and Akutsu, H., *50W super high-pressure sodium lamp.* Illum. Eng. Inst. Japan Conference, paper no. 61 (1983).

Oomen, J.J.C. and Rouwendal, J.W., *High pressure sodium lamp seals based on rare earth aluminates.* High Temperature Lamp Chemistry (Zubler ed.). The Electrochemical Society, Proc. vol. 85-2, pp. 291–312 (1985).

Oomen, J.J.C., Rouwendal, J.W. and Moosdijk, W. van de, Private communication (1985).

Oppenheim, U.P. and Even, U., *Infrared properties of sapphire at elevated temperatures.* J. Opt. Soc. Am., vol. 52, pp. 1078-1079 (1962).

Os, J. van, *Openbare verlichting en dimervaringen.* (Public lighting and dimming experiences.) Elektrotechniek, vol. 60, pp. 475-479 (1982).

Osteen, M.M., *Color improvement of high-pressure sodium vapor lamps by pulsed operation.* U.S. Patent no. 4.137.484 (1979).

Otani, K. and Suzuki, R., *Effects of the polycrystalline alumina arc tube transparency on the efficacy of the high pressure sodium lamps.* J. Light & Vis. Env., vol. 3, no. 2, pp. 18-23 (1979).

Otani, K., Kawahara, K., Watanabe, K. and Tsuchihashi, M., *A high pressure sodium lamp with the improved color rendition.* J. Light & Vis. Env., vol. 4, no. 1, pp. 24-27 (1980).

Otani, K., Watanabe, K. and Tsuchihashi, M., *The effect of the xenon gas on the high pressure sodium arc.* J. Light & Vis. Env., vol. 5, no.2, pp. 1-6 (1981).

Otani, K., Kawahara, K., Watanabe, K. and Tsuchihashi, M., *A high pressure sodium lamp with improved color rendition.* J. Illum. Eng. Soc., vol. 11, pp. 231-240 (1982).

Otani, K., *The convection phenomena in the high pressure sodium lamp.* J. Light & Vis. Env., vol. 7, no. 2, pp. 1-6 (1983).

Ozaki, N. *Temperature distribution of the high-pressure sodium vapour discharge plasma.* J. Quant. Spectrosc. Radiat. Transfer, vol. 11, pp. 1111-1123 (1971a).

Ozaki, N., *Resonance radiations from high-pressure sodium plasma.* J. Quant. Spectrosc. Radiat. Transfer, vol. 11, pp. 1463-1473 (1971b).

Ozaki, N., *Luminous efficiency of the high-pressure sodium lamp.* J. Appl. Phys., vol. 42, pp. 3171-3175 (1971c).

Page, R.B., Denbigh, P.L., Parott, M.A. and Molesdale, P.D., *Future trends in high pressure sodium lamps.* CIBS National Lighting Conference, pp. CB1-5 (1980).

Parrott, M.A., *Translucent ceramics as lamp envelopes.* Light. Res. & Technol., vol. 6, pp. 19-23 (1974).

Patterson, P.L., *Modulated spectra of a.c. arc lamps.* J. Opt. Soc. Am., vol. 62, pp. 627-633 (1972).

Peelen, J.G.J., *Light transmission of sintered alumina.* Philips Tech. Rev., vol. 36, pp. 47-52 (1976).

Peelen, J.G.J., *Alumina: sintering and optical properties.* Ph. D. thesis, Technical University of Eindhoven (1977).

Penning, F.M., *Über den Einfluss sehr geringer Beimischungen auf die Zündspannung der Edelgase* (Influence of very small additives on the breakdown of noble gases). Z. Phys., vol. 46, pp. 335-348 (1929).

Perrin-Lagarde, D. and Lennuier, R., *Mesure expérimentale du coefficient d'absorption de la vapeur de mercure dans la région de la raie de résonance* $\lambda = 2537$ Å (Experimental determination of the absorption coefficient of mercury vapour near the 2537 Å resonance line). C.R. Acad. Sc. Paris, vol. 247 B, pp. 1020-1023 (1972).

Pichler, G., Milošević, S., Veža, D. and Beuc, R., *Diffuse bands in the visible absorption spectra of dense alkali vapours.* J. Phys. B: Atom. Molec. Phys., vol. 16, pp. 4619-4631 (1983).

Popov, K.G. and Ruzov, V.P., *Collision broadening of the sodium D_1 line, ($\lambda = 589.6$ nm).* Opt. Spectrosc. (USSR), vol. 48, pp. 372-373 (1980).

Prud'homme van Reine, P.R., Private communication (1981, 1982).

Prud'homme van Reine, P.R., *The influence of impurities and dopes on the thermophysical stability of translucent Al_2O_3.* Proc. 12th Int. Conf. on the 'Science of Ceramics', Saint-Vincent, Italy, (Vincenzini ed.), Science of Ceramics, vol. 12, pp. 741-749 (1983).

Rautenberg, T.H. and Johnson, P.D., *Time-dependent plasma temperature measurements of the high-pressure sodium arc.* J. Appl. Phys., vol. 48, pp. 2270-2273 (1977).

Reck, G.P., Takebe, H. and Mead, C.A., *Theory of resonance absorption line shapes in monatomic gases.* Phys. Rev., vol. 137A, pp. 683-698 (1965).

Redwood, M., *Mechanical waveguides.* Pergamon, Oxford (1960).

Reints Bok, W., *Study of the wall corrosion in high pressure sodium lamps with improved colour rendition.* Extended abstracts Spring Meeting Toronto, The Electrochemical Society, vol. 85-1, pp. 567-568 (1985).

Reiser, P.A. and Wyner, E.F., *Use of the peak shifts of the 3S-3P sodium resonance lines for the analysis of high-pressure sodium lamps.* J. Appl. Phys., vol. 57, pp. 1623-1631 (1985).

Rhodes, W.H., *Sintered yttria for lamp envelopes.* 2nd Int. Symp. on Incoherent Light Sources, Enschede, Netherlands, pp. 103-104 (1979).

Rickman, J.D., *High-pressure sodium lamp arc tube temperature and alumina sublimation.* J. Appl. Phys., vol. 48, pp. 3733-3738 (1977).

Rigden, S.A.R., *High-pressure sodium vapour discharge lamps.* G.E.C. Journal, vol. 32, p. 37 (1965).

Rigden, S.A.R., Heath, B. and Whiscombe, J.B., *Electric discharge lamps.* US Patent 3.473.071 (1969).

Rijswick, M. van, Private communication (1981).

Rittner, E.S., Ahlert, R.H. and Rutledge, W.C., *Studies on the mechanism of operation of the L Cathode.* I. J. Appl. Phys., vol. 28, pp. 156-166 (1957).

Rockwood, S.D., *Elastic and inelastic cross sections for electron – Hg scattering from Hg transport data.* Phys. Rev. A, vol. 8, pp. 2348-2358 (1973).

Ross, J.F., *Ceramic bonding.* US Patent 3.281.309 (1966).

Rothe, D.E., *Radiative ion-electron recombination in a sodium-seeded plasma*. J. Quant. Spectrosc. Radiat. Transfer, vol. 9, pp. 49-62 (1969).

Rozenboom, J., *Electronic ballast for gas discharge lamps*. Third Int. Symp. on the Science and Technology of Light Sources, Toulouse, France (1983).

Ruff, H.R., *Street lighting – application engineering requirements of lamps and lanterns*. Public Lighting, vol. 120, pp. 232-248 (1963).

Rutledge, W.C. and Rittner, E.A., *Studies on the mechanism of operation of the L cathode. II*. J. Appl. Phys., vol. 28, pp. 167-173 (1957).

Saito, N., Akutsu, H., Kondo, K. and Gion, H., *Electrode design of high pressure sodium lamps*. Illum. Eng. Inst. Japan Conference (1978).

Schäfer, R. and Stormberg, H.P., *Investigations on the fundamental longitudinal acoustic resonance of high pressure discharge lamps*. J. Appl. Phys., vol. 53, pp. 3476-3480 (1982).

Schlejen, J. and Woerdman, J.P., Private communication (1985).

Schmidt, K., *Metal vapor lamps*. US Patent 2.971.110 (1961).

Schmidt, K., *Emission characteristics of 'high-pressure' sodium discharges*. Bull. American Physical Society, vol. 8, p. 58 (1963a).

Schmidt, K., *Radiation characteristics of 'high pressure' alkali metal discharges*. Proc. 6th Int. Conf. on Phenomena in Ionized Gases, Paris (Hubert and Cremien-Alcan Eds), Part III, pp. 323-330 (1963b).

Schmidt, K., *Parameters of the high pressure sodium discharge column*. Proc. 7th Int. Conf. on Phenomena in Ion. Gases, Beograd (Gradevinska Knjiga), vol. I, pp. 654-658 (1965).

Schmidt, K., *High-pressure sodium vapor lamp*. US Patent 3.248.590 (1966).

Schmidt, K., *High-pressure saturated vapor sodium lamp containing mercury*. US Patent 3.384.798 (1968).

Scholz, F.S., *Characteristics of acoustical resonance in discharge lamps*. Illum. Eng., vol. 65, pp. 713-717 (1970).

Shao-Zhong, Li, Tai-Ming, Zhou and Zu-Quan, Cai, *Analysis of factors affecting color rendition of HPS lamps*. High Temperature Lamp Chemistry (Zubler ed.). The Electrochemical Society, Proc. vol. 85-2, pp. 313-322 (1985).

Shkarofsky, I.P., Johnston, T.W. and Bachynski, M.P., *The particle kinetics of plasmas*. Addison-Wesley (1966).

Smyser, W.E. and Speros, D.M., *Discharge lamp thermionic cathode containing emission material*. U.S. Patent 3.708.710 (1970).

Spitzer, L., *Physics of fully ionized gases* (Chapter 5). Interscience, New York (1956).

Stoer, G., *Lighting for floriculture, Hensbroek, The Netherlands*. International Lighting Review, vol. 34, pp. 100-103 (1983).

Stormberg, H.P., *Line broadening and radiative transport in high-pressure mercury discharges with NaI and TlI as additives*. J. Appl. Phys., vol. 51, pp. 1963-1969 (1980).

Strickler, S.D. and Stewart, A.B., *Radial and azimuthal standing sound waves in a glow discharge*. Phys. Rev. Letters, vol. 11, pp. 527-529 (1963).

Strok, J.M., *New low wattage family of high-pressure sodium lamps*. IES conference, USA (1980).

Stull, D.R. and Prophet, H., *JANAF Thermochemical Tables, 2nd edition*, Nat. Bur. Standards, Washington (1971).

Svehla, R.A., *Estimated viscosities and thermal conductivities of gases at high temperatures*. NASA Technical Report R-132 (1962).

Szudy, J. and Baylis, W.E., *Unified Franck-Condon treatment of pressure broadening of spectral lines*. J. Quant. Spectrosc. Radiat. Transfer, vol. 15, pp. 641-688 (1975).

Teh-Sen Jen, Hoyaux, M.F. and Frost, L.S., *A new spectroscopic method of high pressure arc diagnostics*. J. Quant. Spectrosc. Radiat. Transfer, vol. 9, pp. 487-498 (1969).

Tielemans, P.A.W., *Measurement of electrode losses in high-pressure gas-discharge lamps*. Conf. on Electrode Phenomena in Gas Discharges, paper 6.10, Bucharest (1974).

Tielemans, P. and Oostvogels, F., *Electrode temperatures in high pressure gas discharge lamps*. Philips J. Res., vol. 38, pp. 214-223 (1983).

Tiemeyer, F., Private communication (1977).

Tiernan, R.J. and Shinn, D.B., *Sodium reactions with high pressure sodium arc tube material*. High Temperature Lamp Chemistry (Zubler ed.). The Electrochemical Society, Proc. vol. 85-2, pp. 251-260 (1985).

Tol, T. and Vrijer, B. de, *Electric discharge lamp comprising container of densely sintered aluminium oxide*. US Patent 3.726.582 (1970).

Touloukian, Y.S. and DeWitt, D.P., *Thermophysical properties of matter*, vol. 7: *Thermal radiative properties*. IFI./Plenum, New York (1970).

Trigt, C. van and Laren, J.B. van, *On radiative transfer in gas discharges*. J. Phys. D: Appl. Phys., vol. 6, pp. 1247-1252 (1973).

Unglert, M.C., *The need for high-pressure sodium ballast classification*. Light. Design. & Appl., pp. 29-32 (March 1982).

Unglert, M.C. and Kane, R.M., *The interaction of high-pressure sodium lamps, ballasts and luminaires*. Light. Design & Appl., pp. 43-47 (November 1980).

Unsöld, A., *Physik der Sternatmosphären* (Physics of stellar atmospheres). Springer-Verlag Berlin/New York (1968).

Vargaftik, N.B. and Voshchinin, A.A., *Experimental study of the thermal conductivity of sodium and potassium as vapors*. Teplofiz. Vys. Temp., vol. 5, pp. 802-811 (1967).

Verderber, R.R. and Morse, O., *Energy savings with solid-state ballasted high-pressure sodium lamps*. Light. Design & Appl., vol. 12, pp. 34-40 (January 1982).

Veža, D., Rukavina, J., Movre, M., Vujnovic, V. and Pichler, G., *A triplet satellite band in the very far blue wing of the self-broadened sodium D-lines*. Opt. Comm., vol. 34, pp. 77-80 (1980).

320

Vliet, J.A.J.M. van and Nederhand, B.R.P., *Time-dependent measurements and model calculations of high-pressure Na-Xe discharges.* (29th Gaseous Electronics Conf., Cleveland) Bull. Am. Phys. Soc., vol. 22, p. 195 (1977).

Vliet, J.A.J.M. van and Broerse, P.H., *Electric device provided with a metal vapor discharge lamp.* U.S. Patent 4.117.371 (1977).

Vliet, J.A.J.M. van and Jacobs, C.A.J., *New developments in high-pressure sodium discharge lamps.* CIBS National Lighting Conference, Canterbury, pp. CA1-7 (1980).

Vliet, J.A.J.M. van and Groot, J.J. de, *High-pressure sodium discharge lamps.* IEE Proc. Pt A, vol. 128, pp. 415-441 (1981).

Vliet, J.A.J.M. van and Groot, J.J. de, *Einfluss der Wandtemperatur auf die Energiebilanz und die Spektrale Energieverteilung der Hochdruck-Natrium Entladung* (Influence of wall temperature on the energy balance and spectral energy distribution of high-pressure sodium lamps). Frühjahrstagung der Deutsche Physikalische Gesellschaft, Wurzburg (1982).

Vliet, J.A.J.M. van, *Ignition of gas discharge lamps.* Third Int. Symp. on the Science and Technology of Light Sources, Toulouse, France (1983).

Vliet, J.A.J.M. van and Groot, J.J. de, *Influence of wall temperature on luminous efficacy of high-pressure sodium lamps.* Third Int. Symp. on the Science and Technology of Light Sources, Toulouse, France (1983).

Vrijer, B. de, *Hogedruknatriumlamp* (High-pressure sodium lamp). Electrotechniek, vol. 43, pp. 512-514 (1965); Industrie Elektrik & Elektronik, vol. 11, pp. 401-403 (1966).

Vrugt, P.J., *Compatability of alumina (sapphire) and sodium in the temperature range of 1075-1525 K.* High Temperature Lamp Chemistry (Zubler ed.). The Electrochemical Society, Proc. vol. 85-2, pp. 237-250 (1985).

Ward, P.C., *In-situ evaluation of high pressure sodium (HPS) lamp capacity for gettering hydrogen.* J. Illum. Eng. Soc., vol. 9, pp. 194-196 (1980).

Ward, P.C., *Estimation of HPS lamp voltage rise as a function of ballast characteristic curve parameters.* J. Illum. Eng. Soc., vol. 13, pp. 157-161 (1983).

Waszink, J.H., *A non-equilibrium calculation on an optically thick sodium discharge.* J. Phys. D: Appl. Phys., vol. 6, pp. 1000-1006 (1973).

Waszink, J.H., *Spectroscopic measurements on a high-pressure Na-Xe discharge and comparison with a nonequilibrium calculation.* J. Appl. Phys., vol. 46, pp. 3139-3145 (1975).

Waszink, J.H. and Flinsenberg, H.J., *Determination of the electron density in a high-pressure Na-Xe discharge from the profile of a Stark-broadened spectral line.* J. Appl. Phys., vol. 49, pp. 3792-3795 (1978).

Watanabe, K., *Resonance broadening of the sodium D lines.* Phys. Rev., vol. 59, pp. 151-153 (1941).

Watanabe, K., Saito, M. and Tsuchinashi, M., *Emission and evaporation characteristics of $Ba_{1.8}Sr_{0.2}CaWO_6$ emitters.* J. Light & Vis. Env., vol. 1, no. 1, pp. 13-17 (1977).

Waymouth, J.F., *Electric discharge lamps*. M.I.T. Press, Cambridge, Massachusetts (1971).

Waymouth, J.F., *An elementary arc model of the high pressure sodium lamp*. J. Illum. Eng. Soc., vol. 6, pp. 131-140 (1977).

Waymouth, J.F. and Wyner, E.F., *Analysis of factors affecting efficacy of high pressure sodium lamps*. J. Illum. Eng. Soc., vol. 10, pp. 237-244 (1981).

Waymouth, J.F., *Analysis of cathode-spot behaviour in high-pressure discharge lamps*. J. Light & Vis. Env., vol. 6, no. 2, pp. 5-16 (1982).

Wharmby, D.O., *Scientific aspects of the high-pressure sodium lamp*. IEE Proc. Pt A, vol. 127, pp. 165-172 (1980).

Wharmby, D.O., *Energy balance of high-pressure sodium discharges under controlled vapour conditions*. J. Phys. D: Appl. Phys., vol. 17, pp. 367-378 (1984).

Wiese, W.L., Smith, M.W. and Miles, B.M., *Atomic transition probabilities*. National Stand. Ref. Data Ser., Nat. Bur. Stand. (USA) 22, vol. 2, pp. 2-8 (1969).

Williams, C.E., *Circuits for starting high-pressure sodium lamps*. J. Sci. & Technol., vol. 37, pp. 35-40 (1970).

With, G. de, Vrugt, P.J. and Ven, A.J.C. van de, *Sodium corrosion resistance of translucent alumina: effect of additives and sintering conditions*. J. Mater. Sci., vol. 20, pp. 1215-1221 (1985).

Witting, H.L., *Acoustic resonances in cylindrical high-pressure arc discharges*. J. Appl. Phys., vol. 49, pp. 2680-2683 (1978).

Woerdman, J.P. and Groot, J.J. de, *The Na$_2$ singlet and triplet absorption spectrum observed in a high-pressure sodium discharge*. Chem. Phys. Lett., vol. 80, pp. 220-224 (1981).

Woerdman, J.P. and Groot, J.J. de, *Emission and absorption spectroscopy of high pressure sodium discharges*. ACS Symposium Series, no. 179. Metal Bonding and Interactions in High Temperature Systems (Gole and Stwalley Eds), pp. 33-41 (1982a).

Woerdman, J.P. and Groot, J.J. de, *Identification of NaAr and NaXe satellite bands: a test for calculations of the $B^2\Sigma^+$ potential*. J. Chem. Phys., vol. 76, pp. 5653-5654 (1982b).

Woerdman, J.P., Schlejen, J., Korving, J., Hemert, M.C. van, Groot, J.J. de and Hal, R.P.M. van, *Analysis of satellite and undulation structure in the spectrum of Na + Hg continuum emission*. J. Phys. B: At. Mol. Phys., vol. 18, pp. 4205-4221 (1985).

Wyner, E.F. and Maya, J., *Sodium dimer emission observed in arc discharge afterglow*. 29th Gaseous Electr. Conf. U.S.A. (1976); Sylvania Laboratory Report/LR-65 (1977).

Wyner, E.F., *Total emittance measurement of arc tubes in high-pressure sodium lamps*. 2nd Int. Symp. on Incoherent Light Sources, Enschede, Netherlands, pp. 28-29 (1979a).

Wyner, E.F., *Electrolysis of sodium through alumina arc tubes*. J. Illum. Eng. Soc., vol. 8, pp. 166-173 (1979b).

Wyner, E.F., Gungle, W.C., Zack, A. and Cohen, S., *Incandescent socket adapter for low wattage mercury and sodium lamps*. J. Illum. Eng. Soc., vol. 11, pp. 79-84 (1982).

Yu-Min, Chien, *On the shifts of self-reversed maxima of the sodium resonance radiation*. J. Appl. Phys., vol. 51, pp. 2965-2968 (1980).

Yu-Min, Chien and Lin-Tang Chen, *Resonance radiation power from a high-pressure sodium lamp*. J. Phys. D: Appl. Phys., vol. 15, pp. 775-783 (1982).

Zakharov, A.I., Persiantsev, I.G., Piśmennyi, V.D., Rodin, A.V. and Starostin, A.N., *Dynamics of streamers directed by an outer electrode for a breakdown in xenon*. Teplofiz. Vys. Temp., vol. 12, pp. 252-258 (1974).

Zemke, W.T., Verma, K.K., Vu, T. and Stwalley, W.C., *An investigation of radiative transition probabilities for the $A^1\Sigma_u^+ - X^1\Sigma_g^+$ bands of Na_2*. J. Molec. Spectrosc., vol. 85, pp. 150-176 (1981).

Zollweg, R.J., *Molecular radiation in high pressure sodium arc lamps containing mercury or cadmium*. Proc. 11th Int. Conf. on Phenomena in Ionized Gases, Prague, p. 402 (1973).

Zollweg, R.J., *Physics and chemistry of high pressure sodium lamps*. ACS Symposium Series, no. 179. Metal Bonding and Interactions in High Temperature Systems (Gole and Stwalley Eds), pp. 407-420 (1982).

Zollweg, R.J., *The alumina arc tube wall temperature of high pressure sodium lamps*. J. Illum. Eng. Soc., vol. 12, pp. 275-281 (1983).

Zollweg, R.J. and Kussmaul, K.L., *Non-destructive monitoring of high pressure arcs*. Light. Res. and Technol., vol. 15, pp. 179-184 (1983).

Index